Space Science and Astronomy Theatre

———— ✦✦✦✦✦ ————

Margaret Boone Rappaport
and
Christopher J. Corbally

ARCHWAY
PUBLISHING

The information on science and social science in this book, Space Science and
Astronomy Theatre, is non-fiction, and based on the expertise of the two authors.
As with all science, updates to scientific information may be forthcoming in the
future. The twelve scripts in this volume are works of fiction, and while some
of them use historical events and persons, the other characters and the action
in the scripts are fictitious and the joint work of the two authors' imagination.

Scripture taken from the King James Version of the Bible.

Interior Graphics/Art Credit: NASA/JPL-Caltech; Stellarium; Yikrazuul

Archway Publishing books may be ordered through booksellers or by contacting:

Archway Publishing
1663 Liberty Drive
Bloomington, IN 47403
www.archwaypublishing.com
1 (888) 242-5904

ISBN: 978-1-4808-4844-3 (sc)
ISBN: 978-1-4808-4845-0 (e)

Library of Congress Control Number: 2017909278

Print information available on the last page.

Archway Publishing rev. date: 07/20/2017

PART 1. Contents

FOREWORD

This book is a perfect antidote for anyone who ever thought that astronomy and space science were austere and difficult subjects, designed to be taken seriously and appreciated from afar, like marble sculptures in a museum. In this book, the anthropologist-astronomer team of Margaret Boone Rappaport and Chris Corbally has breathed life into the science by turning it into 12 short scripts of character-driven drama and comedy.

Space Science and Astronomy Theatre is aimed at young people contemplating a career in science. The material is suitable for middle school up to college age. As students and their teachers present and discuss these two-character scenes, they will get a complete snapshot of the astronomical life—excitement and frustration, mind-bending concepts, and a rich cultural backdrop. I suspect the scripts could also play a valuable role in creating interest and awareness among students who have no intention of majoring in science.

Scenes range from 160,000 years in the past to 2000 years in the future. Roles range from a Chinese court astronomer to a Cockney "wise guy." Minimal props and costumes are required—this is intimate theater, ideal for a small setting. The authors are true to the science but they are also playful. The tone is breezy and light, with nods to the popular culture alongside classical references. Everything you need to get started is here—scientific contexts, directions and scripts, lesson plans, and materials to help with evaluation.

"Let's put on a show!" This was the ethos of classic MGM musicals starring the young Judy Garland and Mickey Rooney. That infectious spirit permeates this book. Staging a play is the ultimate learner-centered way to teach and engage students, so this book should help illuminate the path to a science career for a new generation.

Chris Impey
University Distinguished Professor
Associate Dean
Department of Astronomy
University of Arizona
Tucson, Arizona

PART 1

SPACE SCIENCE AND ASTRONOMY ROLES IN TODAY'S SOCIETY

A. Tomorrow's Workforce

Welcome to this introduction to a unique, new instructional tool to interest middle school, high school, college, and perhaps younger students, in the variety of available career paths in space science and astronomy. While astronomers remain at the core of basic scientific research in space science, and astronauts are the most visible practitioners of space science, there are many other specialties that students can pursue to take part in humanity's off-world adventures in space—from engineering to accounting, graphic arts to public relations, and biology to meteorology.

The Space Science and Astronomy Script Packages found in part 3 of this volume will guide teachers and instructors in planning short dramatic presentations whose key concepts and lessons are in science. The discussions of science and technology careers will help parents discuss possible career paths with their children. We hope to provide you with useful information and a special new curriculum to increase your students' interest in space science and to show them ways that they can take part in some of the most exciting projects on Earth and in space.

We have devised a dozen two-actor dramatic presentations for use by young adults from high school to freshmen in college, and perhaps middle school students in general science classes. Each script package includes a five-minute script for two characters, one female and one male, along with stage directions. (Teachers can easily re-craft for two males or two females.) Accompanying each script are materials for instructors that transform the short dramatic production into a learning unit. There are keywords and study questions for use before the script and discussion questions for use after the script. We provide background information in the form of "lessons in history and culture," and "lessons in space science and astronomy" for each script. We also provide suggestions for simple props and costume pieces, along with recommendations for background slides that can be prepared with PowerPoint or a similar program.

This volume is not primarily for the parents and teachers of a fully committed high school senior or college freshman who knows he or she is destined for a career in science or engineering. However, those students may find the scripts amusing and informative, and the scripts may help them consider which sub-field to enter. The scripts provide an opportunity for the committed science student to lead discussions, answer particularly difficult questions, and guide the review of keywords. If given leadership roles, students who are already bound for a doctorate in science can maintain a high level of interest in science, and other students will learn from them.

The scripts are designed for students who may have given light to moderate consideration to pursuing a science major and a science career. It is also geared towards students who are captivated by space projects and astronomy but do not know how to articulate their interests with available career paths in the space agencies and astronomical research facilities. For these students—who include many females and minority students—this volume will help them learn about the wide variety of science roles available to them, while they gain some basic science education about phenomena like supernovae, star formation, giant molecular clouds, and the search for habitable exoplanets. We encourage the reader to dip into one or two of the scripts in part 3, to get a feel for them, before continuing with the orientation below in part 1 and their use in part 2.

B. The Space Science and Astronomy Script: The Fascination of Astronomy, the Utility of Drama, and the Interplay of Religion

A Matrix of Interdisciplinary Studies

This book lies within a matrix of many scientific disciplines. In our dozen Space Science and Astronomy Script Packages, we trace processes and structures that begin with the Big Bang, then follow the emergence of molecular clouds, stars, and galaxies, and finally, we focus on human interaction in scripts from periods of time as long ago as 160,000 years, to a time as far in the future as A.D. 3054. In many ways, the pathos of human struggle remains the same, although the cultures vary widely from one epoch to another, from one script to another.

It is the thread of human sentience that interests us most, those special qualities that set our species apart and make it unique. Members of the genus, species, and subspecies *Homo sapiens sapiens* are the only ones who presently exhibit a profound sense of self-awareness and an ability to solve problems through creative social action and symbolically transmitted patterns. In science, religion, and art, human beings find their highest expression, as well as a certain responsibility for the welfare of other species and the environment of the Earth. Human responsibilities are as great as human capacities.

Broad Appeal

After performing all of the five-minute scripts in this volume for a variety of audiences (high school students, college students, dinner guests, club members, and conference attendees), we have concluded that the scripts in this book have a curiously broad appeal. Each of the scripts takes drama, human social participation, and instruction in science and culture, and then blends them into an effective mix that "works." People laugh. People learn. People consider new ideas.

The reactions from audiences are initially curiosity and amusement, such as when they first see the costumes for the A.D. 3054 supernova sighting (the authors show these and other costumes on the website, http://thehumansentienceproject.org). Then, they become interested in the topics in science and culture, and ask questions. Most individuals have not experienced dramatic productions in the venues where we have performed—in a school, at a planetarium, at an academic conference, at a club meeting for church members, in homes, and even in a public library. It is an unusual occurrence, and the near-uniqueness of the Space Science and Astronomy Script is part of the appeal of a scripts event.

Size of the Venue for a Scripts Event

Why a surprised and amused reaction occurs in response to a scripts event is worth exploring, but perhaps not too deeply. We believe that the scripts are effective because most dramatic performances are usually not so close, either in proximity or in the fact that onlookers can interact with the performers afterward, or they can take turns and become performers, themselves. Most people are familiar with dramatic productions that take place on a large stage, and the onlookers are far removed from the action. For a scripts event, a smaller location is preferable. We recommend that, where possible, the scripts retain a small-venue, intimate appeal, like a neighborhood theatre, a dinner theatre, or a theatre-in-the-round.

Informality

A second factor explaining the appeal of the scripts may be their informality. The authors have performed the scripts only by reading the scripts from a sheaf of pages, never by memorizing and speaking the lines. As scientists, both authors are accustomed to formal presentations at academic meetings. In fact, it was their confrontation of the *un*desirability of a formal presentation that initially gave rise to the concept of the Space Science and Astronomy Script. The very first script was conceived as a novel way to get across ideas about the interrelation of astronomy, culture, and art, for adult colleagues at a conference on the "Inspiration of Astronomical Phenomena," at the Hayden Planetarium in New York City. Over the next six months, other scripts were performed in private homes and at public gatherings to a variety of audiences. Part 2(A) describes the evaluation of three scripts presented to high school students, who, early in our planning, had become the focus of our educational project. We then provide the results of our evaluation of scripts in several different college classes in astronomy.

It is not necessary for a scripts event to be formal. In fact, a scripts event may benefit from informality, and so scripts are perfectly suited to a small classroom setting. For example, two students can first perform a set of roles, and then another two students can do the same. Indeed, the notion of switching the gender of the two characters in a script—one, a man, and the other, a woman—and allowing students to play counter-types (with appropriately re-named characters), is first suggested in the introduction to Script Package #9. However, this can be done with any of the scripts, or, two males can play the roles, or two females. Names of characters are easy to change.

If students want to present these scripts for a larger audience in, for example, a school assembly, they may want to memorize lines. If they decide to craft a three-act play using one of the scripts as the kernel of a longer narrative, with more characters, that approach will also work in a school auditorium, on a large stage. Indeed, it would be a valuable project in creative writing for some students to write a longer script, and for drama students to perform it. Students oriented to the sciences could lead the discussion of the science questions afterward, and a student who favors social studies could lead the discussion of the Lessons in History and Culture.

Versatility

We found that the scripts are extremely versatile—surprisingly so—and can be both instructive and enjoyable at various levels of sophistication, while still conveying information on science and culture. Also, they can afterward lead to fruitful discussion of the very latest in space science topics. In many ways, the scripts "speak for themselves" when they are performed, because they deal with universal emotions, conflicts, and problems, and each one has both tension and resolution. The excitement of scientific discovery in Scripts #2 and #11 cannot be lost on anyone in the audience. In Script #7, the quandary of Seer, who is torn between staying home with the woman he loves, and going far from home to find food and a warmer place to live for his band, cannot be dismissed by any onlooker. He is in a jam and his people are in trouble, and his familiarity with "star maps" gives him the confidence to leave home and follow a different kind of map.

The Utility of Drama

We are reminded by a character who is a teacher in Script #5 that confrontation and heavy-handed lessons are usually less effective than lessons delivered in a more intriguing and circuitous manner. For those of us in the sciences, we know full well that science is fun. It can be tedious, frustrating, routine, and repetitive, but the thrill of discovery at the end makes the enterprise rewarding. Science may be the ultimate exercise in "delayed gratification," so it is worthwhile for a student to determine if he or she has the patience, skepticism, orientation to detail, and ability to self-correct to make a good scientist, engineer, or technician. Can a fascination with astronauts, space, and the stars translate into an achievable career? The ranks of astronomers and astronauts are small and positions, few. Can an interest in space morph into a suitable career in a telescope's machine shop or engineering lab? It is best to find out.

The scripts, themselves, introduce a student to the nature of the work involved in scientific research. Is it for them? Do they have the temperament? Thrilling events occur infrequently, and it is important for students to ask themselves if they are good at planning and executing long-term projects, and if it is gratifying to help with them. Is the nature of the work sufficiently engaging that feedback is not necessary each and every day? Can they work alone? Are they good observers? If they have scientific research as a goal, do they have the mathematical and geometric ability that lie at the heart of modeling natural phenomena and mining large databases? And finally, if they aim for a professional degree in science, are they willing to teach it, too? Almost all scientists teach science to others at one time or another.

For high school seniors and freshman college students, these questions can be broached gently and with humor through drama. Why? How? Drama helps them visualize "real people" coping with the real problems of equipment and interpretation of data. Drama calls upon people to be open and expressive of their ideas because the performers are seemingly so open.

Proven Effectiveness

Our evaluation in a classroom setting of three Space Science and Astronomy Script Packages, including a review of keywords and study questions, showed that just over half of a senior class of fifty-six students at a predominantly Hispanic high school in the American Southwest, had an improved orientation to a science major or science career. The most change was not seen among students who were committed to graduating and following a science degree path, but other students, who were not yet committed to a science career.

We improved the performances that followed the evaluation, and we changed the worksheet given to students before the presentations (worksheets with keywords, questions for research before the scripts event, and discussion questions afterwards). We conducted several evaluations of scripts in introductory college classes that had many freshmen and sophomores. Again, scripts were improved and additional lessons were learned about how the scripts worked to encourage openness to a career in science. In particular, we examined the effects of: an existing commitment to a major; number of college credits already earned; and whether the students performed the scripts after they were performed for them. (See the evaluation findings and Best Practices in part 2(A) of this volume.)

The Fascination of Astronomy

We would be remiss if we failed to mention the inherent fascination of the stars as a source of appeal for the script productions. Two of the scripts address this interest more directly than the others, and those scripts are the two that include early *Homo sapiens* in Africa as the characters (Scripts #7 and #10). In both brief stories, the audience can see early humans noticing the positions of the stars in the night sky, and benefitting from that interest. We suggest that humans have been mesmerized by the night sky for many thousands of years, because it was a part of the natural world that changed through the seasons, and thus became an early source of prediction about weather, animals, and plants. Knowing the cycles of game reproduction and seed formation meant that there was some way to predict food resources. Unsurprisingly, both scripts with early humans concern food, although from very different perspectives.

Human fascination with the stars and planets now has a new focus, and it has more to do with the future of humanity than its past. When we look up now, we see places where we know that humans will someday travel, live, go to school, grow crops, and have children. These will be places where we work, and some of that work is illustrated in the scripts: for example, Scripts #3, #5, and #12, where we see a future anthropologist, a future college teacher, and a future miner, respectively. These scripts are in keeping with our goal of illustrating the wide variety of occupational roles that someone could fill who is interested in space science and astronomy.

The Interplay of Religion

It is because of fascination, fears, and an urge to control the uncontrollable—all emotions illustrated in the scripts—that we must mention the interplay of the stars and sky with religion. One of the authors is not only an astronomer, but a Catholic priest, and his insights about stars and religious beliefs are provided here to help bridge the gap between the two juggernauts of modern thinking: science and religion. We include religious thought in several of the scripts, for example, Script #7 on early humans, whose protagonist, named "Seer," exclaims, "The spirits are restless tonight, Em. Can you hear them calling in the wind? Can you see them speaking to us in the twinkling stars?" Another example is Script #4, whose characters mention several Inca gods, and the changes in religion required by the consolidation of the Inca Empire.

The thoughts on religion and the stars that follow here come primarily from individuals who see no conflict between science and religion, and they are provided not as an effort to proselytize, but as an important part of our scripts-based profile of how humans think, which is the focus of our work in the interdisciplinary field of "Science and Religion" (Corbally & Rappaport 2013; Rappaport & Corbally 2015).

We ask: Have you ever looked at the stars on a really dark night, far away from the city lights? The number of stars that one can see is overwhelming. Yet, the number of stars seen by the naked eye is only around five thousand, and most of these stars are in our very local neighborhood of the Milky Way Galaxy. From our research on other galaxies and clusters of galaxies in the universe, we know that there are literally billions and billions of stars. Ten times the grains of sand on all the beaches on Earth roughly tally up to the total number of stars we think are in the observable universe. The cosmos is immense and its impact on us on a dark, starry night can bring us a profound feeling of awe and wonder. A believer in a Deity and humans with a sense of the supernatural can join those from ancient times, singing, "The heavens declare the glory of God" (Psalm 19, 1). The stars and the supernatural have been interwoven in human thinking for thousands of years, and we hazard a guess that they will be for a long time to come.

The vastness of the universe can frighten us. If it does, it is comforting to remember that science shows us that the vastness of both space and time has a life-oriented outcome. Huge amounts of hydrogen and helium gas, with small quantities of associated dust particles, coalesced to form generation after generation of stars. These succeeding generations of stars synthesized an increasing storehouse of elements in the universe that are essential to life: carbon, oxygen, nitrogen, and trace elements like phosphorous and potassium. Without this vast stellar factory for the elements (featured in Script #8), working through billions of years, we would not have an Earth to provide the elements out of which we are made. Many scholars interested in evolution suggest that there are tendencies in the cosmos toward higher levels of organizational complexity (e.g., Davies 1995: 103 ff). Science has given us the possibility, but not the proof, that the universe is in some sense centered on producing life.

People of faith will add that this story of the evolution of life in the universe is an affirmation of God's inexplicable and very personal Love, which has been weaving the human story through immense spans of time. Again, the ancients have anticipated this insight: "When I consider thy heavens... the moon and the stars... What is man, that thou art mindful of him...? For thou hast made him a little lower than the angels, and hast crowned him with glory and honour" (Psalm 8, 3-5).

An astronomer not only deals with vastness, but with the microscopic. The physical processes generating radiation in the cores of stars—radiation that arrives at our telescopes sometimes after a very long journey—occur at the size-level of the atom. It is in understanding the interplay of radiation with atoms that we can understand what stars are all about, or penetrate what is happening in the giant clouds of gas and molecules, such as the Orion Nebula featured in Script #8. These microscopic events in the universe, whose characteristics are formalized in the laws of physics, are given to us as unique, and to the believer they represent a God-given milieu for scientific interactions. They lead us to the ever-active, innerness of God's loving presence in creation. As the paleontologist and priest, Teilhard de Chardin (2002), wrote, "At the heart of Matter is a heart of the World, the heart of God."

The study of astronomy and physics gives glimpses of how the universe works. It brings joy to both the professional and amateur astronomer. The authors want to share this joy through the scripts and lessons in this book. The scripts, like poetry, can convey so much more than mere words on paper or lectures in the classroom. By using this book, we invite parents, teachers, and students alike to enter more fully into the joy of the universe, of which they are not simply a part, but they embody the whole of its evolution.

C. Space Science and Astronomy Now and in the Future

1. Manpower needs; issues for women and minorities

Space science is a broad field that encompasses astronomy, but also a great many other disciplines, too. While astronomers are important in generating basic knowledge about the cosmos, others can follow their vision of an involvement with space by pursuing complementary courses of study and obtaining degrees and experience in a multitude of fields. Figures on occupations within "space science" are not easy to estimate, because participants are trained in so many different fields and serve at many different levels in a variety of organizations.

The term "space science" has grown in use, while the career opportunities in the more basic research sciences of astronomy, astrobiology, and astrogeology remain limited. When one asks whether there are sufficient numbers of students for these specialized fields and enough training programs for them, the answer is generally, "Yes." Indeed, the fields are, and have been, a bit crowded. It is common for astronomers to begin their careers with a series of low paid, post-doctoral positions, until they obtain permanent jobs, usually in higher education or government. Early post-doctoral positions can require repeated re-location to different facilities, and so can be difficult on families and pocketbooks. More serious in terms of planning educational programs is the risk that astronomers will leave the field altogether because it is so competitive. Some do.

Working Astronomers

Within the broad field of space science, astronomers are at the core of basic scientific research. However, the National Research Council's Decadal Survey (NRC 2010) also emphasizes the social utility of the field.

> Astronomy offers a high return on investment for the United States, attracting young people to science and technology careers and providing the kind of education and training that can help solve major societal challenges... (NRC 2010: 103)

It is difficult to determine exactly how many astronomers are working today in the United States, or worldwide, since that category is determined primarily by a definition of individuals who have a certain education—a doctorate in astronomy or astrophysics. However, an individual trained in astronomy may have switched to an applied field in physics and be working in another occupation, or may have switched out of astronomy altogether, and thus not be counted. Others trained in closely related fields, for example, astrogeologists and astrobiologists, may be classified by their organizations as "astronomers." Engineers working in fields closely related to space science are also sometimes members of astronomical organizations. All these factors make the counting of astronomers difficult.

The American Astronomical Society (AAS), the main professional organization for astronomers in North America, reported in 2005 that "astronomy is a relatively small field, with only about 7,000 professional astronomers in North America" (AAS Web Site 2017). The AAS also reported an annual rate of about 50-100 permanent job openings for astronomers in North America, and about 240 Ph.D. graduates in astronomy/astrophysics per year. Most of the new job openings for astronomers are postdoctoral or otherwise impermanent positions, and many young astronomers transfer out of academia after doing a couple of postdocs. The International Astronomical Union (IAU) has just over 10,000 individual members from 79 different countries (IAU Web Site 2017).

This is not a large workforce, but astronomers enjoy the advantage of knowing each other and collaborating more closely than members of a larger group. When jobs in astronomy are examined by discipline, field, and location, the NRC Decadal Survey reports that the top "disciplines" are Observational Astronomy–Optical (as opposed to observation using other wavelengths of light), and Theory. The main "field" is Stellar Astronomy, followed by Galaxies and Clusters. The "location" of astronomers is dominated by the Research University category, followed by the Government Laboratory (NRC 2010: 117). It should be noted that these figures are from 2009, and since that time, interest in Exoplanets has grown enormously. That category is not broken out in the 2009 data. As in any discipline, interest in astronomy subfields can emerge and grow very rapidly. This is often dependent on advances in instrumentation and/or new data sources, such as the Kepler field data on exoplanets that is featured in two of the scripts in this volume (Scripts #9 and #11).

A U.S. Census Bureau report on "Detailed Occupations and Median Earnings: 2008," lists "aerospace engineers" and "astronomers and physicists" among the most highly paid occupations, with median earnings of $91,230 and $89,190, respectively (2008: 3). When only men's wages are considered, the median earnings are $92,650 and $98,360, respectively (2008: 7). Among "Top-Ranking Science Salaries," the salary of an "experienced senior physicist" is $106,716, and the salary of an "experienced astronomer" is $87,534 as of December 2005 ("The Salaries of Scientists by Discipline" online, from Jupiter Scientific Publishing). However, from their own experience, individual astronomers might well question these figures.

Space Scientists

There is no category of "Space Scientist" in the Census reports on U.S. occupations. The main occupational categories for work on space projects at the National Aeronautics and Space Administration (NASA) or the European Space Agency (ESA) are in the engineering fields. While these large space agencies fund many grants and contracts on which astronomers serve, this remains indirect funding.

NASA lists the following job categories as comprising sixty percent of their positions at the "Professional, Engineering and Scientific" level: Accounting, Aerospace Engineering, Biology, Computer Engineering, Computer Science, General Engineering, and Meteorology ("NASA Occupations," NASA Web Site 2017). These are the professional categories that are most likely to be found useful by today's space agencies and private corporations. However, we are aware that we write at a time of enormous change in space science, when many government functions are being transferred to private enterprise. From every indication, this trend will continue, and occupations in space science may change with its application and commercialization.

When the U.S. Census Bureau examined STEM (science, technology, engineering, and mathematics) occupations, the individuals who work in three occupational codes most clearly include professionals in "space science"—aerospace engineers, astronomers and physicists, and atmospheric and space scientists (Table 1). Yet, these occupational categories cover many others who are not in "space science" and who are involved in aircraft design and testing, physics research (pure and applied), and meteorologists not involved in space projects.

	2010 SOC Code	Numbers 2011	Percent of STEM Workforce	Percent Female	Percent White, Not Hispanic
Aerospace engineers	17-2011	124,902	1.7	11.3	77.8
Astronomers and physicists	19-2010	11,331	0.2	19.7	71.3
Atmospheric and space scientists	19-2021	8,407	0.1	16.6	82.9

Table 1. Selected Data from "Disparities in STEM Employment by Sex, Race, and Hispanic Origin. American Community Survey Reports, September 2013," Table 3. SOC Codes, from "Crosswalk of Full List of 2010 Census Detailed Occupation Codes to EEO Occupation Codes."

At the present, there is no way to compile figures on all people working in "space science," or the broader category of individuals working in non-professional positions for space science projects, such as technicians, programmers, and support personnel.

Women and Minorities

The STEM workforce in the three occupational categories detailed in Table 1 is 70-80 percent white and predominantly male, although women and minorities are making some advances in these fields. However, their low representation remains a cause for concern. The NRC Decadal Survey reports the following.

> Black Americans, Hispanic Americans, and
> Native Americans constitute 27 percent of the
> U.S. population. By all measures they are seriously
> underrepresented among professional astronomers ...
> this cohort accounts for only 4 percent of astronomy
> Ph.D.s awarded in the United States and 3 percent of
> faculty members, and yet even these small fractions
> represent growth (NRC 2010: 125-6).

The same survey points to some progress with respect to women in astronomy, although they continue to be underrepresented in comparison to the general population.

> The fraction of astronomy graduate students that
> are women has increased from a quarter to a third
> over the past decade, and the fraction gaining Ph.D.s
> and occupying assistant and associate professor
> positions is also a quarter. However, only 11 percent
> of full professors are women... (NRC 2010: 128).

To address the disparities in STEM occupations by sex, race, and Hispanic origin, a large number of secondary school and college programs have been developed to encourage greater interest in science by women and minorities, including mentoring, group-study, and scholarship programs (Koenig 2009: 1386-87). It is well known that women often switch out of science courses in favor of humanities curricula, but less well known is the fact that women also transfer into the sciences from the humanities (Mervis 2014: 125-6). Clearly, the process of encouraging young women and minorities to enter science fields is complicated. The study by Mervis states, "Few would argue with the need to make the STEM workforce more diverse. But the studies provide no clear guidance on where to focus attention" (2014: 126). Similarly, the programs for encouraging minorities in STEM fields have "Good intentions, scant data" (Koenig 2009:1387).

With respect to astronomy, specifically, the NRC Decadal Survey reports that "women were once as underrepresented in professional astronomy as minorities are today... Now, there is ongoing progress toward parity..." (NRC 2010: 128). However, minorities remain woefully underrepresented, producing a small fraction of doctoral level degrees (NRC 2010: 125-6). Gender bias remains in both the United States and in Europe (Fohlmeister & Helling 2014).

Needed Knowledge Base

Therefore, not only are figures on professionals and technicians in "space science"—even professional astronomers—difficult to estimate, data on what programs encourage young women and minorities in space science fields or any STEM field, for that matter, are sorely lacking, and more research is needed. In light of this absence of understanding, we were glad to be able to evaluate three of our scripts with students in a largely Hispanic high school in the Southwest, and we were gratified by the results. A summary of the findings of the evaluation appear as the first section in part 2 of this volume. We evaluated Script #9 for two Introductory Astronomy classes at a community college in the Pacific Northwest, and we assessed yet other scripts less systematically before a wide range of audience types, primarily adults with young-adult children.

2. Overview of roles in space science and astronomy, for teachers and parents

In space science and astronomy, it appears that there is truly something for every interest and each skill level. This section provides a selective overview of roles in some of the most recent growth fields within astronomy and space science. In one case—energy applications—there is little growth now, but we anticipate more growth in the future. There are specialties that link and cross-cut many of the fields we mention, for example, instrumentation, nanoscience, and optics. Our overview of research and applied roles is not meant to be exhaustive, but to give teachers, parents, and administrators the information needed to help them guide young people.

Each of the science activities mentioned in this overview of roles relies on a legion of electronics technicians, engineers (especially computer and mechanical engineers), computer programmers, as well as machinists in a facility's workshops and labs. A student need not achieve a Ph.D. to be involved in many of the fields and subfields mentioned below. Many of these technical-level positions are well compensated and provide secure employment, with benefits.

The scripts in this book were designed to showcase a certain amount of interdisciplinary work, now and in the future. Therefore, they can be useful in encouraging students in chemistry, biology, and geology, as well. Not only does "space science" imply a great deal of collaboration between scientists and engineers, but astronomy, itself, now implies large and growing subfields that involve colleagues in other science disciplines.

For example, researchers in molecular chemistry collaborate with astronomers to understand the interstellar medium. The study of interactions within molecular clouds—well represented in Scripts #4 and #8—is a bona fide subfield in both astronomy and physical chemistry. Planetary scientists and biologists collaborate in the field of astrobiology. At the same time, many researchers in biology, chemistry, and physics are requiring a growing knowledge of statistics and computing methods. For example, astronomers model the atmospheres of both planets and stars. They even model the interactions of entire galaxies, giving, for example, a prospective view of our Milky Way and the Andromeda Galaxy colliding some four billion years in the future. Through citizen science, which is well adapted to astronomical research, many scientists and educators are working with experts in machine learning to use input provided by human beings to develop smarter computers. Research scientists in all fields are beginning to use expert programs and AIs (artificial intelligences)—which are featured in Scripts #9 and

#11—to develop and test theories, classification schemes, and the behavior of physical, chemical, and biological processes. One of the authors of this book has helped to develop an expert program to classify stellar spectra. He and a colleague have found that the program can filter out some kinds of "noise" (random fluctuations) even better than humans!

Administrative positions should be highlighted in our overview of roles in space science and astronomy. Our scripts feature project managers and the manager of a commercial business. Research scientists must also develop some administrative skills, themselves, because much of the funding for space science and astronomy research projects comes from external government and foundation sources, and periodic reporting and proposal writing are critical. The line between pure and applied research is increasingly fuzzy, and as a result, professional astronomers can change fields temporarily and take on some applied work for a period.

Scientists rely every day on a bevy of management roles and many types of administrative assistance to track hours and expenses, arrange for travel, write and negotiate contracts, manage their facilities, and coordinate public relations. Both the pure and applied sides of space science require some expertise in data handling and analysis, project management, and public communication and education activities, all of which require broader training than the physical sciences. Behind every research scientist are support personnel whose knowledge of basic science can only be an added qualification for their work on space science and astronomy projects.

The first three research roles below are from the "priority science objectives" chosen by the NRC Decadal Survey for 2012 to 2021. These are (1) searching for the first stars, galaxies, and black holes, (2) seeking nearby habitable planets, and (3) advancing our understanding of the fundamental physics of the universe (NRC 2010: 2). An astronomy and astrophysics Decadal Survey is conducted regularly by the National Research Council and is highly influential in framing future research policy.

a. Research roles for professional astronomers

Extragalactic Astronomy and Galactic Evolution

What came first: the chicken or the egg? When this question is asked in the context of galactic evolution, the question involves whether stars or galaxies were primary. It is a very fundamental question because the answer requires explanation of how and why the early universe was not uniform in structure but very slightly clumpy, and, how the first stars could manage to form and become the progenitors of the hundreds of billions of stars that were gathered into galaxies. How do the different shapes and configurations of galaxies, with massive black holes at their centers, fit into our account of the evolution of galaxies up to the present? These are heady questions!

Extragalactic astronomers are concerned with questions that focus beyond our own Milky Way Galaxy and yet, involve it, too. Their approach can be theoretical or observational, or a combination of the two. The theoretical side relies on modeling galaxies and their interactions, using computer simulations. The observational side calls on very large telescopes that are operational now, or being planned, and on innovative space facilities that will detect, for example, gravitational waves.

A person who looks at the universe as a whole is called a cosmologist. Much of the cosmologist's work is theoretical, but it also has a strong observational component, for example, the study of a particular type of supernova, known as "Ia," to determine the distances and redshifts of their galaxies. Different types of supernovae are featured in Scripts #1, #2, and #3, but some of the methods for investigating them are the same. Supernovae greatly impact the evolution of structure within a galaxy. For example, they influence the shape of a galaxy's spiral arms. Therefore, an extragalactic astronomer must be aware of the stars that are a part of galaxies, as well as all other galactic components and influences.

Extragalactic astronomers are not featured in any of the scripts, but we do focus upon the field in part 2(C) of this volume, "How to Develop Additional Astronomy Script Ideas." Furthermore, the work of extragalactic astronomers forms a foundation for much of the content of the scripts. This will become clear in the introductions to the individual astronomy script packages. The budding extragalactic astronomer may well have already discovered the citizen-science project that is part of the "Zooniverse" online science series. If so, he or she is already familiar with "Galaxy Zoo," an interactive online project that is steadily classifying the most distant galaxies of the universe. We hope it is proving enjoyable!

Planetary Science

Are we alone? The universe is vast and it makes us wonder whether we are the only intelligent life in it. The planetary scientist helps to provide part of the answer by investigating how our own solar system came to be. Was it a usual product of star formation, or was our solar system somehow unusual? A complex picture is emerging that includes all the solar-system "debris" such as asteroids, comets, and meteoroids left over from our star's—the Sun's—formation. The planetary scientist asks about the origins of our Moon and the moons around other planets in our solar system, and the planets and moons around other stars.

The finding of large Jupiter-class exoplanets very close to their host stars has challenged the standard picture of planetary formation, which had traditionally seen large planets as remaining distant from their stars. This finding has led to understanding the possible migration inward of "Hot Jupiters" within their planetary systems. Today's planetary scientist must be an exoplanetary scientist, as well.

There is still so much that we do not know about our solar system and for which the observing tools of the planetary scientist are essential. There are all sizes of telescopes from medium to large, and all kinds of probes from orbiters to landers. These can visit planets, moons, asteroids, and comets and gather data in all parts of the electromagnetic spectrum, from xrays to the radio.

Planetary scientists sometimes team with planetary geologists (astrogeologists) to investigate the composition of other planets, their satellites, and asteroids. In Script #12, we see the practical consequences of learning about the composition of asteroids. Their precious metals may be a source of revenue, but they can also yield water and metals for the construction of space-based telescopes, stations, and vehicles. This avoids the costly delivery of materials off the Earth.

Script #5 assumes that our Moon has become like a space station and is a good location for a university branch. Script #3 looks forward to a more distant time when humans routinely have research bases on the Jovian moons like Europa, and new communication technologies to link them almost instantly to Earth. The characters in Script #9 are nearing a habitable exoplanet when a communication comes from an unexpected direction. Communication also seems to have been a problem in Script #12, when a miner becomes stranded on an asteroid during mining operations. In the action of all these scripts, the planetary scientist's knowledge has been an essential foundation.

The New Physics

What *don't* we see in the universe? The answer from current astrophysics and observational cosmology is a disturbing, "about ninety-five percent of everything!" Cosmologists tell us that around sixty-eight percent of the known universe seems to be dark energy and about twenty-seven percent is dark matter. This leaves a mere five percent as matter that we are able to see and understand. "Dark energy" is what we call the mysterious force that is accelerating the rate of expansion of the universe. This increase in the rate of expansion began about a third of the way back in time since the Big Bang, at 13.8 billion years ago. We see the acceleration's effect on the expansion of galaxies away from each other, but we do not know what is causing it.

What we call "dark matter" is causing the outer parts of galaxies to rotate faster than they should, given the matter that we can detect.

In other words, there seems to be "missing mass" (the original term for dark matter). Gravitational lensing, where a foreground cluster of galaxies bends the light from a very distant galaxy, also reveals dark matter. We cannot see dark matter, although it certainly exists, nor can we detect it directly, like we can detect usual matter. Furthermore, we cannot see black holes, for example, the massive black hole that is at the center of the Milky Way Galaxy and features in Script #6. However, we know that black holes exist because their considerable gravity speeds up stars in their vicinity and gives them away. Dark matter is similarly given away by the speed at which the outer parts of galaxies rotate. Dark energy is revealed by the faster-than-expected expansion of the most distant galaxies.

The mysteries concerning dark energy and dark matter call for people trained in physics and mathematics, who work in laboratories with particle accelerators and the sub-atomic particles they produce. Knowledge gained in experimental physics should lead to better answers to fundamental questions that astronomers have about the behavior of galaxies. Astronomers will soon be working with enormous quantities of data received from a new generation of telescopes planned for Earth and near-space. The Large Synoptic Survey Telescope (LSST), scheduled to begin full science operation in 2023, will be a particularly powerful tool. The LSST program will need astronomers and engineers with a variety of specialties, and the technicians and support personnel to help achieve LSST's science goals.

Scripts #1, #2, #3, and #6 feature supernovae and their products—black holes and pulsars. The explosions of the more massive stars result in black holes, and the less massive stars result in pulsars. The characters in these scripts are all involved in the development of "the new physics," by recording new observations of these phenomena. While dark matter and dark energy are not explicitly treated in any of the scripts, when characters mention "galaxies" and the evolution of stars since the Big Bang, they imply a foundation from the science of dark energy and dark matter.

Stars and Brown Dwarfs

When is a star not a star? Stellar astronomers can now answer, "When it is a brown dwarf." Some products of star formation in the giant clouds of gas and dust in our galaxy are so low in mass that they cannot sustain nuclear fusion in their cores, although there might be a little at the start of their lives. These bodies are now called "brown dwarfs." They were first postulated by theoretical stellar astronomers and were observed beginning in the 1980s. Today, brown dwarfs are proving to be a rich source of knowledge about the so-called "weird chemistry" occurring in their relatively cool atmospheres. For example, the atmospheres of brown dwarfs are known to precipitate silicates and iron, and so "rain hot rocks and molten iron." The insights gained from the study of brown dwarfs extend to the next group of even-lower mass objects called, "the Jupiter class of planets."

An exciting field for stellar astronomers is the study of "outliers" in galaxies, i.e., stars outside the "disk" and "bulge" of galaxies, but inside their "halos." Investigations began with faint stars in the Milky Way galaxy, but interest has spread, thanks to large telescopes, to include stars in nearby galaxies. Streams of stars in the less dense halos of galaxies are found to have very similar compositions, but are unlike any other group of stars inside the host galaxies. Astronomers have concluded that these streams of stars are relics from dwarf galaxies that merged in the past with the large host galaxies.

There are many other puzzles about stars, although stars have been the most studied objects in the history of astronomy. One of the authors is working with colleagues to understand why a certain group of stars—the Lambda Boötis stars—differs from normal stars in chemical composition. This group of stars was first identified in 1943, but the mystery behind their peculiar composition continues, so wish them luck!

Stellar astronomy provides a background to many of the instructional script packages. Scripts #1, #2, and #3 feature the observation of exploding stars, or stars "going supernova." Script #6 features rare stars which not only speed through the halo of the Milky Way, but even escape it completely, given enough velocity. The "star factory" in Script #8 includes important ideas about how stars are formed initially from giant molecular clouds into clusters (for more on molecular clouds, see Script #4). Eventually, most stars drift away or are ejected from their birth-cluster and become free-moving like our Sun and the stars hosting planets in Scripts #9 and #11.

Space Science and Biology

How do we eat in space? The answer at present is that we carry supplies of food with us. However, for very long trips in space the weight of provisions would be prohibitive. We need the help of biologists and agronomists, like those who run the University of Arizona's Controlled Environment Agriculture Center (CEAC) and who conduct experiments in prototype greenhouses that could operate on other planets or on long interstellar voyages. Food has been grown hydroponically on the space station for years. Growing food onboard should be the solution, but much more research is needed on the best plants and the best environments, and how to control the subsystems that ensure plants survive—like colonies of bacteria, mold, and insects.

Astronomers and biologists who are interested in life in other solar systems join forces to understand the fundamental questions on the origin of life. They ask: What were the conditions under which life on Earth started, and can we duplicate these in the laboratory? How might life start in other solar systems, and on exoplanets around other stars? We learn much from "extremophiles" found thriving in the extremes of temperature, pressure, and environment on Earth.

Among scientists who are looking into these fascinating questions are astrobiologists, as in Script #11. Script #9 features an agronomist and greenhouse manager. The training for astrobiologists includes astronomy and biology, as well as an ability to conduct computer simulations. Some astrobiologists have opportunities to go to exotic locales, and even to explore the deep trenches in the oceans or to retrieve ice core samples from the arctic regions of the Earth.

Data Mining

The universe is vast and has the potential to create for us an "avalanche" of data. Surveys like the Sloan Digital Sky Survey have already produced more information than professional astronomers can easily analyze. The Gaia spacecraft is continuing to send massive streams of data. It is making a three-dimensional catalog of a billion astronomical objects, discovering hundreds of thousands of new objects and observing them all, seventy times, during a five-year period. While all these data are being interpreted, the LSST (Large Synoptic Survey Telescope), from 2023 onwards, will be giving us fifteen terabytes of new data each night.

"Mining" the data produced by these powerful instruments involves developing software to find regularities, irregularities, patterns, and special cases. Data mining for these surveys can be a very complex but exciting computational endeavor. The channels of data to be correlated are numerous and the interpretations of data are often not obvious. In the face of complications, astronomers like to say that there is a "fuzziness" built into the universe. (See the section on "The New Physics," above.)

In astronomy, as in many other fields, data mining is becoming an essential task. It engages both the engineers who can make faster and cheaper computers with more storage, and the software experts who can write the increasingly sophisticated algorithms to process the data. Therefore, a person might start in astronomy and migrate to any commercial, research, or government job that includes data mining. We can draw an analogy from automotive engineering, where Formula 1 racing cars have 120 sensors that produce more figures per race than words we speak in a lifetime! That kind of data flow—and much, much more—needs a creative data mining approach to make the substantial investments in instrumentation worthwhile.

Studies of the Milky Way (Scripts #4, #5, and #6) will benefit enormously from mining the data streams resulting from the Gaia and LSST projects. Even the choice of the asteroid to be mined in Script #12 (to net the greatest revenue from precious metals) will need some data mining, too, as will the choice of stars with exoplanets, which are featured in Scripts #9 and #11.

b. Interdisciplinary and applied roles in space science

The first two applied roles below (in aerospace engineering and computer engineering) are featured because they are prominent in the professional and technical ranks of the large space agencies, NASA and ESA. The next four roles (in applied astronomy, public education, energy applications, and defense applications) are mentioned because they currently include applied astronomers. There are few astronomers in these fields now, but we believe they are areas of future growth. The last two applied roles (in archaeoastronomy, and in science and theology) are listed because they are strikingly interdisciplinary, and are growing with the participation of professional astronomers.

Aerospace Engineering

Aerospace engineering is one of the principal areas for professional positions at NASA and ESA. An aerospace engineer designs, builds, and tests aircraft for operation within Earth's atmosphere and spacecraft for operation in space. An aerospace engineer may work with materials engineers, computer engineers, and physicists to create lightweight designs for onboard propulsion, computer, and life-support systems.

Engineering and computer science are the best career tracks for someone who aspires to a "professional, engineering, and scientific" role at NASA. If one is accustomed to seeing NASA flight directors cover important space launches and exciting Mars rover landings, it is important to remember that the primary professionals behind these events are engineers. Seven specialty areas account for sixty percent of NASA's professional positions, and three involve engineering: aerospace engineering, computer engineering, and general engineering. Other specialty areas include biology, meteorology, accounting, and general computer science. Technical positions (which account for nine percent of NASA's positions) also reflect the nature of the professional-level jobs: electronics, engineering, and meteorology.

Administrative and management positions with NASA and ESA can be hugely rewarding and give someone a sense of involvement with America's and Europe's premiere space agencies. These roles involve budgets, contracts, human resources, public affairs, public education, and information technologies and their application. While NASA and ESA are the big players, private organizations are increasingly sending payloads into space. These include large firms such as Boeing and smaller companies such as SpaceX. They all need aerospace engineers, as well as management and support staff.

Script #6 features a NASA flight director (who can have an engineering background) reporting on an important astronomical event, along with a broadcast media announcer. Script #11 shows an engineer working with an astrobiologist, helping to receive large quantities of data from an orbital telescope and displaying the data in usable form. The "science officer" in Script #9, which takes place in the distant future, probably has skills in engineering and communications, and basic knowledge of spectroscopy (a branch of astronomy), as well as other technical skills and science training. Script #5 includes a near-future technician who works at the docking facilities for rockets landing and taking off from an American moonbase. While his level of science training is not equivalent to the engineer's, it does require substantial knowledge of aerospace systems, and he uses that knowledge in the college classroom we glimpse in the script.

Computer Engineering

Computer engineering in a space science environment involves the integration of networks and systems of hardware and software for specific applications. This may include the design of individual components (such as circuits, microprocessors, sensors, and monitors), but the emphasis is on bringing together a variety of hardware and software in an overall design. Computer engineers are also responsible for testing and evaluating onboard computer and electronic systems for space flight, as well as computer systems related to communications with a home base, life support, and various scientific research and assembly projects.

Computer engineers are an important occupational grouping at NASA. They are indispensable to the construction of aircraft and space craft, and they usually work with a variety of managers and technicians who are part of a team. They are essential to the design, building, commissioning, and operation of today's exquisitely sensitive telescopes and integrating their instrumentation (such as a spectrograph to obtain spectroscopic observations). Along with a crew of technicians, computer engineers ensure the smooth running of vehicles in space, on distant planets and asteroids, and the telescopes that provide data to astronomers.

Teamwork, flexibility, and an ability to fit solutions to problems are critical to the computer engineer's work. We see a computer engineer working in Script #11, to gather biosignature data from an exoplanet. We see two professional astronomers working and monitoring a total of five telescopes in Script #2. Because they are working alone in the middle of the night, they are necessarily performing some of the tasks that a computer engineer might do for a larger project.

Applied Astronomy

Astronomy has traditionally been applied to human problems, for example: (1) the development of calendars which, over the years, have become increasingly consistent with astronomical phenomena, (2) in navigation, especially on the seas and later for aircraft, and (3) for time measurement and standardization. Today, applied astronomy can involve very diverse projects. Indeed, the movement of some professional astronomers to the commercial sector may have good consequences for the field. "Astronomers who go commercial to provide data will be ensuring the long-term future of astronomy," writes Martin Elvis, an astronomer with the Harvard-Smithsonian Center for Astrophysics (2014: 12).

Resource limitations are a constant concern for the relatively modest workforce of professional astronomers worldwide, and the funding of projects and personnel will continue to be an issue in the future. Bachelors and masters degree programs in applied astronomy are being developed, for example, at the University of Michigan, where an interdisciplinary major and minor are offered for students who anticipate careers in teaching, science writing, science journalism, and outreach.

An important applied astronomy endeavor is the removal of "space junk" in the form of old and unused satellites. Their retrieval has become a serious applied astronomy issue for the future use of near-Earth space. Various schemes are in the planning stage for the location and safe disposal of the orbital junk. NASA sponsors an Orbital Debris Program out of the Johnson Space Center in Houston, Texas.

Astronomy can also be applied to the continuing task of understanding solar cycles, anticipating solar flares, and mitigating their effects on the vase electronic and space-based communications networks on, and surrounding our globe. Asteroid mining has made headlines as a potential source of business revenue and several commercial concerns are considering how to begin mining some of the asteroids that come closest to Earth in their orbits around the sun. Script #12 is set far in the future, in the year A.D. 3021, when an asteroid mining company has moved its headquarters to the asteroid belt, where it maintains a commercial interest in mining platinum. Indeed, the S-type asteroid on which the miner is stranded in this script can also contain nickel, cobalt, gold, and rhodium.

Script #7, one of two scripts on early humans in this volume, shows a very early use of applied astronomy. The characters have noticed regularly changing positions of the stars, which helps them to understand where they are and where game may be most plentiful.

Public Education

Professional astronomers often have a significant public education and sometimes, a public relations function. As directors of planetariums and research facilities, or managers of grants and contracts in academia and government, astronomers are sometimes videotaped and interviewed for radio and television programs. This tends to occur when, for example, a bright new comet appears or something unusual occurs like unexpectedly large sunspots that disrupt electronic communications. Neil deGrasse Tyson, Director of the Hayden Planetarium, has hosted a new *Cosmos* series for television. It was pioneered by Carl Sagan's own *Cosmos: A Personal Voyage*, which was broadcast in 1980. *Four Hundred Years of the Telescope* was a documentary in 2009, which featured several different astronomers, including one of the authors of this book.

An increasing number of universities are training STEM (science, technology, engineering, and mathematics) majors in practical leadership and communications skills. For example, the University of Arizona offered a program called, "Getting Students Real World Ready: The Edge," one of which is specifically for STEM students. Students in astronomy who decide to stop at the bachelors or masters levels can find jobs at planetariums, research institutes, universities, and testing facilities in Public Relations. They handle public inquiries, write press releases, manage press conferences and tours, and arrange for press interviews of senior scientists. These middle management positions are often well compensated and provide an avenue for involvement in space science and astronomy.

Writing and reporting on astronomical topics can also provide a rewarding career.

Astronomers trained at all levels can organize and direct the assembly of displays at museums, planetariums, testing facilities, and rocket launch complexes like the Kennedy Space Center in Florida. These venues draw large numbers of visitors each year, and are important for revenue, public relations, and public education. Informing and educating voters about often-costly space enterprises is an activity that helps to ensure continued funding. Astronomers ultimately vet the final details of all exhibits, but they rely on staffs of illustrators, curators, and technicians to provide entertaining and educational displays and shows for the public and for students.

Energy Applications

Research on energy applications is a potential growth area for astronomers, although it is not a significant area of involvement yet. The goal of reducing worldwide dependence on fossil fuels may create conditions that encourage the involvement of more astronomers in energy research. Future activities in space will require planetary scientists to identify off-world propellant sources, because the costs of lifting fuel to Earth orbit will only increase. Development of energy and propulsion systems to carry people and supplies on ever more distant trips in space may require astronomers to work with engineers in designing the energy-related components of space vehicles and stations, especially propulsion, life support, and systems to monitor local conditions (for example, radiation hazards in space). These requirements mainly involve engineers now, as well as many kinds of technicians. Astronomers are likely to have future roles, too.

The plasma physics investigated as part of basic and applied energy research involves the same plasma physics at the centers of stars, and some astronomers "switch over" to energy applications, at least in part. Roger Angel at the University of Arizona is an example of an astronomer who has become interested in solar energy. His specialties include, "concentrating photovoltaic solar energy."

The U.S. Department of Energy funds Energy Frontier Research Centers (EFRCs) throughout the country, primarily at universities and National Laboratories such as Los Alamos and Brookhaven. Many of the research projects have space applications, drawing on solid state physics and optical science to generate the next level of knowledge about energy generation. It is not unusual to see the results of government-funded grants and contracts spin off commercial enterprises.

Script #11 features a science officer of the future who knows quite a bit about warp drive and fusion generators. We can imagine that his knowledge of energy is fundamental to his duties as the science officer of a starship in the year A.D. 2402. In Script #12, the stranded asteroid miner had to be mindful of how much energy, food, and gas supplies he would need before his asteroid came back into conjunction with his home base. He survived only because he knew of his support systems' limitations.

Defense Applications

Many people first become interested in space science because they "want to become an astronaut." It can be a child's earliest expression of interest in space science. Yet, the ranks of astronauts, in the U.S. and in other space agencies abroad, remain quite small. In the past, a good number of astronauts came from the military services, although military service is not now required. Instead, NASA requires "three years of professional related experience, or 1,000 hours of pilot-in-command time in jet aircraft." Interested students should see the NASA web site's pages on "Astronaut Selection," and should keep in mind that the program is very competitive.

Today, the United States military services include an Air Force Space Command, whose mission covers cyberspace as well as space requirements and capabilities. In 2015, Space Command had 38,000 professionals assigned to 134 locations worldwide. Space Command "provides space capabilities for the joint fight through the operational missions of spacelift; position, navigation and timing; satellite communications; missile warning and space control." It also provides "combatant commanders with trained and ready cyber forces," and "designs and acquires all Air Force and most Department of Defense space systems."

Space Command uses many contractors, as well as members of the military services. As in NASA, the ranks are heavily represented by engineers and technicians. Interested students should be directed to the extensive web pages of Space Command, as well as public speeches, and keep in mind that the more security-sensitive programs may not appear. Former head of Space Command, General William L. Shelton, gives a good flavor for the kind of work accomplished, in his speech, "The Value of Space to the Warfighter" (Space Command Web Site, 2014).

Only Script #9 features a military officer, who is serving as the science officer on a starship of the future. However, the students in Script #5 could well represent active members of the military branches who have been assigned to duty stations at a future American moonbase, in A.D. 2101. Young members of the military services often take advantage of the institutions of higher learning where they are assigned, and this could well be the case in Script #5.

Archaeoastronomy

Archaeoastronomy is the investigation of astronomical events as revealed in the prehistoric sites of the cultures of the past. The study often involves astronomers and archaeologists consulting together on remains found in archaeological "digs." Astronomical activities, as revealed in ancient ruins, usually involve a level of labor specialization that signals urbanization. "Astronomers" of the past could well have served as priests, or, as in Script #1, as court astrologers. However, some kind of economic re-distribution had to take place to support a noble, priestly, or scientific class.

Astronomical observations and recording take time and careful study over many years. Only a few people in prehistoric cultures had the luxury of pursuing these activities, which often gave them special powers of ostensible or real prediction. We are often amazed at the level of sophistication of prehistoric cultures when we look back to the archaeological records of the Egyptians, Chinese, Maya, Inca, Arabic cultures, Native Americans (such as Chaco Canyon), and early European cultures (such as Stonehenge). What they learned through observation with their eyes alone, without any telescope, was truly amazing and continues to reveal surprises.

Science and Theology

Astronomers are often asked, "Do you believe there is extraterrestrial life?" or "What was the Star of Bethlehem?" Questions like these are interdisciplinary since their answers, which must inevitably be partial, involve more than astronomy. It is rare that someone trained in astronomy takes up a research or teaching position that is completely interdisciplinary, since that involves education in those other disciplines, but many astronomers are widely read and interested in questions beyond pure science. Their other interests may include theology, and some astronomers take part in meetings and publications with other scientists at the boundaries of science and theology. There are professional organizations who purposefully span science and theology, both in the United States, and internationally. These include the Institute on Religion in an Age of Science (IRAS), the European Society for the Study of Science and Theology (ESSSAT), and the International Society for the Study of Science and Religion (ISSR). These and other organizations have web sites and good resources for instructors, as does The John Templeton Foundation, which funds projects in science and religion.

We have an example of one such astronomer in Script #1. The Chief of the Astronomical Bureau is both an observer of the sky and an interpreter of those observations within the eleventh century system of astrological divination in China. Script #4 involves a harmony between the dark patches of the Milky Way, Inca cosmology and their spiritual world, and agricultural cycles and animal behavior on Earth. In Script #2 there is a discussion of the relevance of astrology. While belief in astrology is anathema to most modern astronomers, they should be aware of its history in pre-scientific astronomy.

PART 2

HOW TO USE THIS BOOK

A. Summary of Best Practices for Using Space Science and Astronomy Script Packages

The following are Best Practices developed from findings of a formal evaluation of the three Supernovae Scripts in Part 3(A), performed and discussed during a single "scripts event" for high school seniors.

Best Practice: The Take-Home Worksheets

Instructors should make the take-home worksheets no longer than the ones provided in this volume. Overly long worksheets are not effective. When possible, assign the hardest questions to committed math/science students in order to keep their interest high.

Best Practice: Time Periods for Script and Scripts Event

A regular class period of 45 to 50 minutes is well suited to the discussion and performance of two scripts. However, the effect of three scripts appears to be worth the added time. Therefore, when there is a longer period of time available (an assembly, a longer class, or a club meeting after school), more scripts are better than fewer.

Best Practice: Choose Three Scripts If Sufficient Time

There was a sharp up-turn in many response scores toward the end of the entire scripts event. Even if the first or second script was accepted with some skepticism, the overwhelming response to a scripts event toward the end of the period was quite positive. For this reason, we conclude that in spite of some fatigue, the package of three scripts came together well.

Best Practice: Prepare for "Students Who Can Go Either Way"

At the high school level, motivated students often show an interest in both science and humanities. Many students can "go either way." The evaluation results suggest that there is also a small proportion of students who are quite committed to science and who seem very likely to follow a science track, but only time will tell if this is true. At the high school level, there is still a great deal of time to switch to other fields. We were pleased to see that among students in our evaluation there were individuals who could "go either way" and who remain open to a science track even if they identify themselves as "mainly humanities," and vice versa.

Best Practice: Humanities Students Are Well Worth Encouraging in the Sciences

Some of the evaluation results suggest that self-identified "mainly humanities" students may have been affected more by the scripts than self-identified "mainly math/science" students. Thirty out of fifty-six of all students (53.6 percent) showed a positive change in their interest in a math/science major or career track by the end of the scripts event. Many of the students who became more interested in science were self-identified as "mainly humanities" students.

Best Practice: Students Cluster into Types, and the Types Are Important for Planning

For the already-committed, high achieving, high school senior who is on a fast track to a science major in college, the scripts may be below their level of experience and they may not respond well to them. A small group of six out of 56 students had scores that declined on Question 4, that is, they responded that the scripts event had a negative effect on their interest in a math/science major and career. Five of this group of six were self-identified "mainly math/science" students and may have already been beyond the level of the material in the exercise. However, the majority of students had scores that changed in a positive direction.

Best Practice: Give Committed Math/Science Students Leadership Roles

We interpret the small group of committed math/science students as already beyond the level of activities pursued at the scripts event. The scripts are not necessarily designed for students who are already committed to a science track, but instead, the middle-range of students and advanced students who are open to an informal learning approach. Teachers may want to prepare the advanced science students differently, either by allowing them to lead the discussion, asking them to report on difficult topics, or suggesting that they can use the scripts as a way to think about a sub-field in which to concentrate.

B. How to Organize Scripts Events for Students

Some preparation and organization of a scripts event will pay big dividends for students and teachers. It is a good idea to gain a general familiarity with all of the scripts, so that good choices can be made on which scripts to perform. The Lessons in Space Science and Astronomy, and Lessons in History and Culture, should fit well with the teacher's required curriculum.

What Is Your Curriculum?

Choices of scripts depend on whether the curriculum is in general science, physics, astronomy, biology, world history, geography, anthropology, creative writing, or drama. These are the courses for which the scripts may be most appropriate in terms of curriculum. The scripts on early humans (Script #7 and #10), as well as the script on biosignatures (Script #11) might find their way into a biology class, as might Script #4 on the Inca. Aspects of the Inca cultivation practices are good examples for both biology and social studies.

Most of the scripts have an interdisciplinary flavor. The Lessons in History and Culture, as well as Lessons in Space Science and Astronomy, which begin each script package, can give teachers good ideas on fitting a script to a curriculum. While the history and culture material is not usually necessary to teach science lessons, it provides a context for the science and may increase student interest. Placing scientific discoveries in a context of human problem-solving tends to increase student enthusiasm for the science. Conversely, a history teacher can use the scripts with the sole purpose of teaching lessons in social studies, including geography. For the social studies teacher, the science lessons can provide interesting, added details, which can be connected to social, political, and economic developments.

Choosing Which Scripts to Perform

The scripts are divided into four major topics in astronomy, with three scripts in each field—twelve scripts in all. Since the scripts, themselves, are brief, two or three scripts from any of the four fields can be successfully combined in a scripts event, which might last about an hour.

Choosing the scripts can be up to the teacher or the students. An introduction to each major astronomy field (Supernovae, Milky Way, Stars, and Planets & Exoplanets) is given as background material for the teacher, as are the Lessons in History and Culture, and Lessons in Space Science and Astronomy, at the beginning of each script package. The three scripts in each major field are listed in chronological order of the events in the scripts.

A teacher of World History might want to select Script #1 and Script #4, as examples of ancient, yet urbanized cultures. The

Chinese of the eleventh century and the Inca of the fifteenth century showed marks of civilization, such as monumental architecture. A teacher of Creative Writing may want to pick out Scripts #3, #9, and #12, as examples of science fiction writing based on real, current events and projections of realistic technologies. It is important to ask students how the script signals that the events are taking place in a future environment. That is a legitimate question for any instructor of Creative Writing.

A teacher of General Science may decide for students to study and perform scripts closest to the present, for example, Script #8 or Script #9. The years A.D. 2032 and 2075 are sufficiently close to the present that generally accurate predictions can be made about, for example, teens in mall arcades, and astronomical research in Hawaii that takes advantage of an orbital telescope.

For those few classes of high school anthropology and the many classes of freshman and sophomore anthropology, Scripts #7 and #10 on early humans in Africa may be an instructive, enjoyable addition to an introductory class or a meeting of an anthropology club. Both scripts are based upon good dates and some of the best archaeological findings. However, we do not know if the early subspecies of *Homo sapiens* were able to converse this well at those dates, so there is a certain amount of "artistic license" in these two scripts.

The terms and topics in the introductions to all of the script packages can be explored further on the internet. Perusing the background material, lessons, keywords, and the scripts themselves, is the best way to decide if a script fits a particular lesson plan and the goals of a certain curriculum. The scripts were designed to be interesting to science, humanities, and social studies students. They are not aimed at just one type of student, and definitely not just science students because some humanities students switch to a science track. Selected speeches in the scripts can be re-written with ease and impunity if the changes are appropriate for a certain audience or if they will craft a better lesson for the students.

Preparing for a Scripts Event

For each script, this book offers a worksheet with "Keywords," and "Questions for Investigation Before the Script," which can be given to students before the script is performed, as a way to prepare. Or, individual questions can be assigned to different students and they can report briefly on them before the script. The authors have checked, and found that simple internet searches will turn up the answers to most of these questions. A verbal introduction to the script's theme, based on the script package's introduction for the teacher, might well accompany the Worksheet(s), or the students can do that research on their own. It is unlikely that students in high school or the first two years of college will be familiar with all of the keywords. A brief introduction to the keywords will ensure that students derive as much understanding from the scripts as possible.

At the time the worksheets are given out, students selected as actors should receive the scripts. They can rehearse before the script performance on their own, or in class, as a way to explore the keywords. At minimum, two students should be selected (or volunteer) to be "actors" for each script. The exception is Script #5, which calls for four actors, although two actors can accomplish the script if one student changes baseball caps with each new speech in the script. In this way, one student plays three characters, with a change of baseball caps. It is quite funny when done this way and lends a lighthearted sense to the college classroom discussion dramatized in this script.

Stage Manager

An instructor can identify a student to serve as a Stage Manager who looks after props, lighting, and microphones, if needed. One of the most valuable roles for the Stage Manager is to be present, with the teacher if possible, when the two or more actors are rehearsing a script. This will help students bring out the humor and drama that is written into the scripts, as well as the scientific sense. The scripts require rehearsal, but always to the point of confidence rather than boredom. Indeed, the scripts' good humor and science instruction is not lost when students who do *not* rehearse are given the opportunity to act out a script. The authors found that "over-practicing" hurt their performances rather than helped them. If the students can commit their parts to memory, so much the better, but we have not found this necessary. The students need a good familiarity with the text, so that they do not stumble over words. Timings are sometimes tricky. This is especially important in conveying the drama of scientific discovery written into the scripts.

The authors found that one of the most important keys to a successful performance of the scripts is sound, and making sure that everyone in the audience can hear. Instructors will know this often requires that student actors be prompted to speak out and project in a "stage voice." Rehearsals are needed so that students speak sufficiently loud and enunciate words well enough, so that the audience can hear and understand. The scripts that require local accents of various types will be a special challenge that may require more practice than the others. If a teacher wants to dispense with the assumed accents, only a little sense will be lost. Mainly, humor will be lost.

Audio-Visual Technicians

The authors found that a set of PowerPoint slides makes a good backdrop for the two performers. The slides can be assigned to one or more students who show a facility with PowerPoint and similar programs, as well as audio-visual aids. Free images are available online for the different kinds of script venues, for example, a telescope control room or a starship "bridge." The internet is a rich resource of images for other scripts, as AV-savvy students will know, but they will need guidance in the choice of these. For Script #9, we selected an image of the bridge of a starship that showed a planet in the portal. This was preferable because it showed that the starship was within a planetary system, not between stars (as they had been for twenty-three years). We excluded images of bridges that showed blue sky outside their portals, since the starship in Script #9 was not that close to a planet. These kinds of decisions will very likely need guidance from a teacher.

The authors attached an audio clip of a rowdy crowd for the "political uprising" in Script #4 on the Inca. For two of the Supernovae Scripts in part 3(A), we used pictures of the night sky produced by the free, planetarium program called *Stellarium* (see Figures 1 and 3). This has a Historical Supernova plugin whose parameters can be changed to show a supernova at any date, position, and brightness. With *Stellarium*'s capabilities, an observer can move from a position on Earth to any position in the solar system, in order to view, for example, Script #3's Cassiopeian supernova from Jupiter's moon, Europa. We showed this view toward the end of the script, when the astronomer finally sees the supernova. It's quite a sight and very exciting!

Most challenging were slides visualizing two events: the Video Arcade Game in Script #8, and the Biosignatures Data that came in slowly, in Script #11. For Script #8, the rapid change of score for someone playing a video game expertly was difficult to re-create. Equally, the incoming top-to-bottom data of two different, but similar, biosignatures (for chlorophyll and hemoglobin) makes Script #11 quite a challenge. At first the authors provided a handout of the molecular structures of the two porphyrin rings to assembled onlookers, on the back of the Program. That sufficed in an informal setting. Students who have talent in using PowerPoint and graphics software will find they can do far better.

Costumes and Props

There is a section on "Simple Costumes and Props" in each Space Science and Astronomy Script Package. These recommendations are based on the authors' experience in finding low-cost alternatives. The costumes and props add a great deal to the all-important "fun element" in performing scripts. We were surprised how little it takes to convey a sense of another time and culture. Our suggestions are ready to be expanded creatively with the involvement of some students—perhaps individuals who can sew. The only sewing that the authors required was stitching around the fake fur pieces used as animal skins for the early humans Scripts #7 and #10. They had a tendency to shed!

While a handful of students should be involved in preparations for the script performances, it is probably best that most of them remain in the audience, having first done the worksheet research, but not having seen the script, itself. The drama of the script has the most impact if it comes as a surprise. The elements of tension, drama, and humor are explained in the introduction to each script for the teacher, just before the script.

Discussion Following Each Script

In each script package, there are "Discussion Questions" for use after the script. More can be added, some dropped, as suits the teacher and students best. However, having a discussion led by the teacher is a vital part of each script, and without discussion, the educational value of the script is much reduced. We have found this to be true even when the scripts are performed in private and relaxed situations, as around a dinner table. The members of the audience always ask questions after the script, and if they are students who are well prepared, their questions creatively explore the science lessons of the script—as we found in our evaluation of scripts for high school students.

Amid these organizational details one element must not be lost. The scripts should be fun for everyone involved: the performers, the support team, the audience, and most importantly, the teacher. This is the very best way to educate: To broaden horizons and to inspire students to become involved in space science and astronomy. We have certainly found this to be true for adults as well as youth who have viewed our performances. Adults are always dumbfounded when they realize that they have "learned something, too."

C. How to Develop Additional Astronomy Script Ideas

Teachers and students can be involved in crafting their own, five-minute space science and astronomy scripts. With appropriate background research, the creation of additional scripts could be a class exercise or a team project, with the result that new lessons in science and culture are learned. The following guidance is given within broad parameters, and teachers will surely modify our script development approach in ways that suit their lesson plans and their students.

The following four steps in creative writing led to the dozen scripts in this book. If implemented, these steps will require some creativity, but are not beyond the ability of teachers who, every day, develop lesson plans, schedules, outlines, take-home worksheets, and exams, as part of their normal routine. All of these teaching aides take enormous creativity, and it is difficult to imagine any teacher in high school or college who could not come up with new space science and astronomy scripts using the following four steps. We give an extended example in the field of Extragalactic Astronomy, and take three ready-made scenarios up to the point of script writing. Students could use these scenarios to develop their own scripts, or students could hold a contest to see who can develop the most interesting and exciting script. Better still, teachers and students can identify other scientific themes on which to base their scripts.

Step One: Identify the Scientific Theme
Step Two: Imagine the Action and Develop a Plot
Step Three: Identify Sources of Conflict in the Script
Step Four: Write the Dialogue and Stage Directions

An Example of the Development of New Script Ideas: Extragalactic Astronomy

Step One: Identify the Scientific Theme

When we began developing ideas for our own scripts, the solid guidance of a professional source on astronomy was critical. The reason for this was simple: Only professional sources can identify the very latest developments in the field and the most exciting events to be anticipated. Fortunately, some science news sources can provide guidance on exciting topics. While the expertise in astronomy for our scripts came from one of the authors, teachers can use yearbooks in astronomy, legitimate and well-vetted science sources on the internet, the *Wall Street Journal*, *New York Times* weekly science section, *Scientific American*, or *Sky and Telescope* monthly, in order to identify the most important news from the field of astronomy. Lead articles signal broad interest, new developments, surprising findings, and therefore, good script topics.

Once we had a list of twelve to fifteen major astronomical events in the past, present, and future, we had the beginnings of our scripts. Below, we list the themes in space science and astronomy that formed the essence—the purpose and the rationale—of our scripts.

Script	Space Science and Astronomy Theme
Script #1	Famous astronomer/astrologer Yang Weide and his assistant discover a "Guest Star" in A.D. 1054. They record it in Chinese historical records, and fashion a likely prognostication to please Emperor Renzong of Song.
Script #2	Two astronomers record the Eta Carinae Supernova of A.D. 2054, which leads eventually to a theory of why and how supernovae occur.
Script #3	Two friends ensure the timely recording of the Rho Cassiopeia Supernova in A.D. 3054, thanks to their use of "sub-space communications," whose signals travel at faster than light speed.

Script #4	The Emperor and his wife recognize Inca spirit gods in the shapes of the Milky Way Galaxy's molecular clouds, and speculate on the connections between heavenly and earthly events.
Script #5	In a college class on "Twentieth Century Mythology," a professor and her students living on Earth's Moon speculate on various methods to locate stars in the Milky Way Galaxy.
Script #6	In a broadcast report on an inter-generational probe to the center of the Milky Way, unexpected failures cause a NASA Mission Flight Director and a journalist surprise and consternation.
Script #7	Early humans recognize that stars occur in regularly changing positions, which can be recorded (an early example of "the external storage of symbols" by early *Homo sapiens*).
Script #8	Two recent high school graduates explore a "star factory" (nebula) in the constellation Orion, using a video arcade game, and speculate on their futures.
Script #9	Crew members of an interstellar ship carrying settlers from Earth to a distant exoplanet, discover their greenhouse systems have failed. They receive an offer of help from an unexpected source and experience First Contact with an alien species.
Script #10	Early humans recognize that a small number of "stars" move in different patterns than the others—thus first discovering planets.
Script #11	An astrobiologist and an engineer discover the first positive biosignature data from an exoplanet near Earth. The findings provide a big surprise.
Script #12	A mining administrator and a miner lament his near brush with death, and apply their knowledge of conjunction and the orbital patterns of asteroids.

The identification of the above twelve astronomical themes, events, and discoveries formed the foundation of our scripts. We then had to imagine the action and develop a plot for each one.

Step Two: Imagine the Action and Develop a Plot

The process of developing the scripts in this volume began with a scientific theme, and then a consideration of possible pairs of roles that might be involved somehow in action involving this theme. The story line begins with an image of the interaction of two people—one male and one female, since the scripts were initially designed to be performed by the authors of this volume. Below, we describe the pairs of roles and their relationship, which ultimately set the boundaries for the interaction between the two characters. The task was to imagine what kinds of exchanges the two might have. For added guidance on character development, teachers and students can consult any number of references.

Script	Pair of Roles
Script #1	A Chinese court astrologer and his female assistant.
Script #2	Two professional astronomers, one helping the other with a project.
Script #3	A professional astronomer on Earth's Moon, and an anthropologist doing research on the Jovian moon, Europa.
Script #4	An Inca emperor who helped to consolidate the Inca Empire, and his wife.
Script #5	A female college teacher at Luna University and her classroom of students on Earth's Moon.
Script #6	A female Mission Flight Director and a male NASA Public Relations Spokesperson.
Script #7	The daughter of an aging head man and her suitor, a male medicine man.
Script #8	High school seniors, both bound for college—a young man and a young woman—who meet at a mall video arcade.
Script #9	A male science officer on a starship, and the ship's female agronomist and greenhouse manager.

Script #10	A male hunter and the young woman he is courting for marriage. She happens to be the progenitor of all living humans in Haplotype "L".
Script #11	Two professionals, one an astrobiologist and the other, an engineer.
Script #12	A female mining administrator and one of the "mine jockeys" (miners) she supervises.

The plot developed sometimes from the beginning of the script, and sometimes it developed from the middle or the end of the script, which is sometimes a surprise or a discovery. The first script, on the observation and recording of Supernova 1054, began simply with the phrase, "Foolish woman!" That set the tone for the interaction between the astronomer/astrologer and his female assistant. The Chinese court at Kaifeng was, after all, in an era in which the Emperor was all powerful, and the aristocrats at court like our noble astrologer, had to be careful. Yang Weide's assistant, who was also a female, was in a far lower position than he, and she knows it. She caters to him and flatters him. He flatters the Emperor, with her good advice.

Other scripts began with an image in the middle. Script #7 began with the carved piece of stone that is a fictional find from an equally fictional Kenya Cave archaeological dig. The one used for the script can be a roughly 6x2x2 inch dimension chunk of plaster-of-paris, painted red. The artifact becomes a kind of map that will show the character Seer how to go to find "the Others" (another group of early humans). During the script, the authors showed an image of the etched stone on a screen, at just the moment that Em takes the etched piece of stone from her animal-skin cloak. Working backward, the action then had to dovetail to that point, and go forward from that point to the end, where there is a resolution of their problem.

Other scripts developed from their endings. The final word "hemoglobin" is spoken at the end of Script #11 on Biosignatures. It is an astounding discovery, and while both professionals half-expect chlorophyll, they never dream they will find evidence of hemoglobin. That has a set of implications all their own, including, "Whom do they notify first? The press or the President?"

Step Three: Identify Sources of Conflict in the Script

Once the characters are imagined, and an image of the scene is firmly established, the dialogue depends on identifying the sources of conflict. That conflict is not always simply between the two characters, but it can be conflict between different ideas about a problem or theory in astronomy, internal conflict in the characters, themselves, conflict with others not in the scene, or conflict arising out of the inherent nature of the situation for human beings (such as waiting for an important, fast-approaching event to occur).

It is important to keep in mind that, for every major field and sub-field in astronomy and for every major news article on space science, there are people here on Earth working and investigating. These people have families, cultures, histories, and conflict. Interesting stories, and therefore scripts, about people always involve conflict. In a five-minute script, the importance of the conflict lies in retaining the interest of the audience. Without conflict, no one can be invested in an outcome, and therefore the onlooker has no reason to be involved. Even a five-minute script has an arc of action: drama that builds, and then resolution and reduction of conflict when the solution to a problem is found. The exception is a largely humorous script, where humor helps to propel interest during otherwise tedious events, such as waiting for a probe to reach the center of the Milky Way Galaxy, as in Script #6.

Identifying the sources of conflict in the story line is the most important task in generating new scripts for scripts. Below, we take a look at the sources of tension for the dozen Space Science and Astronomy Script Packages in this volume.

Script	Sources of Tension
Script #1	Tension between individuals of different sexes and social classes.
	Tension between an expert and his assistant.
	Tension about how to present an essentially bad omen as a good one to the Emperor, and save the astronomer/astrologist's life!
Script #2	Tension about the acceptability of beliefs in astrology, and joking about it.
	Tension about the future: Will a star go supernova tonight?
	Tension inherent to scientific discovery.
Script #3	Tension of time: Will she notify him of the supernova *in time?*
Script #4	Tension between spouses about worshipping older vs. newer gods.
	A ruler's tension about his empire's gains, and his ability to control newly conquered groups of people.
Script #5	Tension about an upcoming class quiz: Will students be prepared?
	Tension, frustration, and humor from misunderstanding information and events.
	Tension from utter surprise: The teacher's reaction to the student from another moonbase.
Script #6	Tension from not knowing about an important event that is fast approaching.
	Tension from very broad humor written into a tense, dramatic piece: What will she say next and how will she say it? Does she know she sounds a little foolish in front of billions of viewers, or not?
Script #7	Tension about leaving a familiar place and finding a new place to live.
	Tension inherent to not understanding a new object, and trying to.

Tension arising from differing opinions about something important.

Script #8   Mock tension between two teens who are accustomed to joking with each other.

Tension between the sexes—boyfriend and girlfriend.

Tension in young people about not knowing their futures.

Script #9   Tension from incomprehension: Why did the AI awake them?

Tension about greenhouse failures and not having enough to eat.

Tension about receiving a signal offering help from the "wrong direction."

Tension about First Contact: What should they do and how should they behave? Are they in danger or not?

Script #10   Tension between a romantic suitor and the woman he admires. Can he woo her and will she agree to marry him?

Script #11   Tension and frustration about equipment failures.

Tension during data transmission, while results are slowly revealed.

Tension of scientific discovery: Are their discoveries valid, and what are the implications of the first biosignature data indicating plant and animal life around another star?

Script #12   Tension from anger at someone for a betrayal.

Tension from inner conflict about a past decision.

Step Four: Write the Dialogue and Stage Directions

The format for writing dialogue varies with advice from different books and expert sources, but it approximates the format given in this book. The dialogue should be appropriate to the two roles chosen for the script, and it should also be consistent with the style of the historical (or future) era, the social positions of the characters, and the relationship of the two people in the script. The best way to determine appropriateness of dramatic language is to experiment. Write lines that seem to be right, but then read them aloud with a friend to see if they sound believable and are easy to speak. The authors omitted more than one "tongue twister" that could not be pronounced easily and might cause a stumble.

Give stage directions only when they are absolutely needed. If the story line involves a significant movement of a character, a gesture, an orientation (to a person or an object, such as a flatscreen), which is important to the sense of the action, then include a stage direction. On a poster board or PowerPoint slide, provide information that sets the scene: Characters, Place, Location (interior or exterior), and Time.

An Example of the Development of New Script Ideas: Extragalactic Astronomy

Here, we explore our approach for scripts development in an area of astronomy that we have not used in any of the dozen scripts in this volume, although we have given summaries of topics within Extragalactic Astronomy in part 1(C) on research roles for astronomers. In spite of the enormous importance of the field, we found that we could not project a time in which humans would have the capability of traveling to other galaxies. The nearest spiral galaxy is Andromeda, which is two and a half million light years distant. We can model Andromeda and the Milky Way colliding billions of years in the future, but we cannot imagine a technology that would take us to this distant galaxy within the foreseeable future. Therefore, intergalactic space voyages were necessarily excluded, and for all practical purposes, so were probes to distant galaxies. We were concerned that we might be confined to scenarios involving complex computations and modelling, and, while computer models of extragalactic evolution are often visually beautiful, intriguing, and increasingly complex, the sources of human conflict we could imagine were mostly obscure—mostly, but not all, as we shall see.

Extragalactic Astronomy is not new, dating from the first observations of "a little cloud" in the constellation of Andromeda, and recognition that some nebulae were "island universes," i.e., other galaxies like our Milky Way, but lying millions of light years distant from us. Development of "standard candles" to assess the distance of far-off galaxies was important. Regularly pulsating stars and later, supernovae provided measures to determine distance based on standard brightness. And then, development of larger and larger telescopes allowed increasingly distant galaxies to be observed and analyzed.

Eventually, astronomers recognized that galaxies were not isolated, but occurred in hierarchies of clusters, and so they began to study their interaction, how they built up, and how they evolved. Our own Milky Way Galaxy is one of fifty-four galaxies in our Local Group—of which Andromeda is the largest. The Large and Small Magellanic Clouds are also in our Local Group. On a scale larger in distance, our Local Group joins other local groups, which together make up the Virgo cluster of galaxies. In turn, the Virgo cluster joins with other galaxy clusters in one of the superclusters of galaxies. Some astronomical work shifted to the mysteries of larger structures of galaxies that appear to be strung out along the surfaces of "voids." We realized that Extragalactic Astronomy could give rise to more script ideas that we imagined at first!

In implementing "Step One: Identify the Scientific Theme," we selected one of the astronomy research priorities detailed in the National Research Council's Decadal Survey (NRC 2010)— Extragalactic Astronomy. This report is one of our two professional astronomy resources, the other again being one of the authors.

In implementing "Step Two: Imagine the Action and Develop a Plot," we were careful to consider pairs of roles that are realistic, and that might involve a certain amount of tension or friction. Here are several pairs of roles that might serve to highlight important research topics in Extragalactic Astronomy.

New Script	Pair of Roles
Script #13	A cosmologist and a computer graphics artist are preparing a presentation for public television on ideas about the distribution of galaxies in the universe and the placement of "voids" where no galaxies are found.
Script #14	An astronomer and an historian are trying to confirm or refute newly found European documentation of as many as five other supernovae occurring around the time of the appearance of SN 1054.
Script #15	An astronomer is guiding his chief engineer in the assembly of an extra-large telescope array that dots an area of near-space on the far side of Earth's Moon, extending outward for a half-million square miles.

The plot—or very short story, in the case of a five-minute script—involving the pairs of roles in Scripts #13, #14, and #15 should necessarily focus on the most active, the most contentious, the most difficult, or most problematical times of their work together. Focusing upon a quiet time is usually not fruitful in developing a dramatic scene, unless the essence of the "quiet action" is fraught with internal conflict or with risk—like the marriage proposal in Script #10, in this volume. The action in the short story should build until there is a resolution, and in the case of short scripts, we have chosen to build often to a "reveal" or an "ah ha moment," when the script ends. In a longer format such as a novel, there would be a resolution phase (or "dénouement" in the language of French dramatists), which would be extended and resolve all outstanding issues introduced during the narrative. In a five-minute script, there is little time for resolution, which is left to the imagination of the audience. Trying to project what happens after the end of the scripts in this volume is a fine exercise for a class in science, as long as it helps to teach additional principles in astronomy science.

Here are several ideas about problems that could evolve between the roles in Scripts #13, #14, and #15. Most important, they must be problems or issues that involve some Lessons in Space Science and Astronomy, which is, after all, the pedagogical goal of the scripts development exercise, although Lessons in History and Culture give even greater depth to a story. In developing potential scenarios, it is important to identify action that is realistic and also very specific. Scenarios that show general problems, but no specifics, are never quite as satisfying to onlookers. Specific details require research, but that research pays off hugely.

New Script   Possible Problem Scenarios

Script #13   Set-up: A cosmologist and a computer graphics artist are preparing a presentation for public television on theories about the distribution of galaxies in the universe, and the placement of "voids" where no galaxies are found.

Possible Problem Scenario: After the LISA (Laser Interferometer Space Antenna) is developed and finally goes online in 2031, the cosmologist works with a graphics artist to illustrate the Triple Black Hole Galaxy (J1502+1115), which had, until LISA went operational, been illustrated with fuzzy diagrams. New data allow gravitational waves within the galaxy to be imaged, but the cosmologist and artist conflict over the use of color. The cosmologist wants to use the traditional colors for hydrogen (red), sulphur (yellow), and oxygen (green), which leaves only blue for the artist to use in imaging the gravitational waves in the Triple Black Hole Galaxy. The only solution is to develop a new color scheme of greys and blues that ripple with the course of the waves.

Script #14   Set-up: An astronomer and an historian are trying to confirm or refute documentation of as many as five supernovae occurring around the time of the appearance of SN 1054.

Possible Problem Scenario: Many historians have searched for sightings of the A.D. 1054 supernova in European records, and there are no confirmed observations of what the Chinese called a "Guest Star" around that time. This script takes place in A.D. 2085, after all of the works in the Vatican Library have been digitized and catalogued. In the script, an historian uses new search criteria to locate astronomical phenomena that might signal the sighting of a supernova that remained visible for approximately two years, as SN 1054 did. The astronomer is brought in to confirm or refute a list of five such possible sightings near the date A.D. 1054. Initially, the astronomer finds five supernovae to be far too many, based on the rate of supernova sightings in other galaxies. Using this extragalactic yardstick, the historian and astronomer debate the list of five possible sightings, and fasten upon one, which is very likely SN 1054. The language used in the documentation is archaic, but there is a clear reference to "yellowish" starlight.

Script #15   Set-up: An astronomer is guiding his chief engineer in the assembly of an extra-large telescope array that dots an area of near-space on the far side of Earth's Moon, extending outward for a half-million square miles.

Possible Problem Scenario: Traditionally, "project scientists" (in this case, an astronomer) and "chief engineers" are often at loggerheads because their goals are different. The scientist wants the best data from instruments, which will function in multiple modes. The engineer wants efficiency, cost effectiveness, and ease of operation and repair. For this A.D. 2160 script, a large-scale, space-based construction project will install instrumentation to study the most distant (and oldest) galaxies. The astronomer and engineer find themselves at odds over the need to filter out natural "resonance" of observational platforms vs. the difficult task of bringing sufficient building materials to a remote location. A solution is found in derailing and slowing an unexpected meteorite with "shepherding rockets." The meteorite is brought into the vicinity of the construction project and mined for extra materials to stabilize observational platforms. The astronomer gets his good data, and the engineer is able to exploit mineral resources already in space, rather than lifting them from Earth.

In implementing, "Step Three: Identify Sources of Conflict in the Script," we identify potential conflict between the pairs of roles we have identified. There are other sources of friction that can be imagined, but these sources of tension lead rather easily to science lessons.

New Script	Sources of Tension
Script #13	A cosmologist and a computer graphics artist are preparing a presentation for public television on theories about the distribution of galaxies in the universe, and the placement of "voids" where no galaxies are found.

- Tension between a professional and a technician.
- Tension between using "traditional" vs. new color schemes.
- Tension and excitement in the creation of a new coloring scheme.

Script #14	An astronomer and an historian are trying to confirm or refute documentation of as many as five supernovae occurring around the time of the appearance of SN 1054.

- Tension between the library staff who "are overworked" and the two researchers, who "want everything."
- Tension between the two experts, when the astronomer concludes (on the basis of extragalactic data) that five supernovae are "too many" for the time period.
- Tension between the two experts and the peer reviewers of the journal considering their paper for publication. The latter do not consider the new library resources sufficient.

Script #15   An astronomer is guiding a chief engineer in the assembly of an extra-large telescope array that dots an area of near-space on the far side of Earth's Moon, extending outward for a half-million square miles.

- Tension between the astronomer's need for data and the engineer's need to keep costs down.
- Tension between the astronomer and engineer, and their project management, when they present their plan to "shepherd" a meteorite into the area for additional construction materials. Management thinks the plan is foolhardy (but in the end, it works).
- Tension when they first test their assembled very large array and evaluate whether they have indeed brought the resonance to a reasonable level.

Step Four: Write the Dialogue and Stage Directions

We shall not write the Scripts #13, #14, and #15, outlined above, although we suggest that students might well take these ideas and craft scripts for their classmates. Indeed, with these three script ideas, a competition could be held for which writing team develops and performs the best script!

D. How to Convert the Scripts to Other Media: Video and Board Games, Home Video, Short Stories, and Novels

We have emphasized that the scripts should be experienced as "fun." This allows their full impact to take hold socially, emotionally, and intellectually. Learning at all three levels is how groups of humans learn, and this learning modality is the advantage that humans have over other primates in mastering a specific task (Dean et al. 2012; Rappaport & Corbally 2015). The impact of the scripts can be increased by extending the topics to video and board games, to home video, short stories, and even novels. Some students will be far more inventive than other students, and sometimes their teachers, in devising ways to encourage this. Below, we venture to give some pointers.

We recognize that some teachers may feel uncertain about encouraging students to "waste their time with video games" and other computer-based projects. To these teachers, we quote an acknowledged expert on the effect of such games, Jane McGonigal, in *Reality is Broken*.

> The truth is this: in today's society, computer and video games are fulfilling *genuine human needs* that the real world is currently unable to satisfy. Games are providing rewards that reality is not. They are teaching and inspiring and engaging us in ways that reality is not. They are bringing us together in ways that reality is not (2011: 4).

With that perspective to consider, we will turn to extending the scripts provided in this volume. Script #8 is set in the context of playing a video arcade game, "Orion the Hunter: Star Factory," which may challenge students to produce some version of that game. Some students will already have written their own video games, and they can be the "game masters" for their classmates and teachers.

Other students are already interested in learning to program in a computer language. That will be an important skill for the rest of their lives. There is some debate about which computer language is the best for beginners. Traditionally, many people began with *BASIC*, but some will now recommend starting with web programming languages like *HTML*, *CSS*, and *JavaScript*. Still others recommend *Python*, which has instruction books aimed at the young beginner and an extension that is specially designed for producing games called *PyGame*. An advantage of *Python* is that it has become a popular tool for data mining and other applications in professional astronomy. The IT (Information Technology) department of a college or a computer science teacher in a high school will be a useful resource for students who want to program games. Internet sources will also be helpful.

If learning a computer language seems daunting to a teacher or to students, there are programs for younger children that are graphically based to such an extent that the details of the computer language rules are not transparent. Students learn the components that go into a video game and the logical sequence of assembling them. Programs include *Scratch, Alice, Kodu, GreenFoot,* Google app inventor, and *Gamemaker.* Some, like *Gamemaker,* have been used at the high school and college level to introduce the logic of programming.

Video games based on the scripts in this volume will require characters that are not readily available in these programs. Character development and drawing is where some students will find a role, and excel. The Art Department of a college, an art teacher in a high school, books on crafting literary characters, and art books will help students to produce characters ranging from realistic to cartoon.

Students can imagine other characters in the scripts provided in this book, name them, draw them, and program their actions. For example, Script #9 mentions the captain of the starship, who never appears in the script. A student could extend the script to waking up the captain and creating a scene around him, as part of a game, or simply a short story. Script #1 mentions the Emperor, but he never appears in the script. A student could extend this story to a programmed video game. One of the authors wrote a short story connecting the two Yang's in Scripts #1 and #2, and won a short story prize in a competition held by a writers club!

A script that lends itself to becoming a board game is #7, which features "a map" of a journey from eastern Africa, northward to warmer parts of that continent. The character "Seer" is encouraged to undertake this trek so he can help his people find more secure food resources. The adventures Seer encounters along the way (such as predators, storms, and volcanic activity) can be selected by the roll of dice. The ultimate prize could be finding another group of hominins to live with, a secure food supply, and returning to Em who consents to be his wife.

Similar adventures could also be the material for a video game that is different from the "shooting game" that springs to life in the video arcade in Script #8, "Orion the Hunter: Star Factory." Perhaps the location for the action in the video game could be changed from the Orion Nebula to the Cygnus-X region or another active star-forming region. A newborn star could be seen "trekking" its way out of danger from supernovae, and from collisions with neighboring stars and black holes in the cluster where it formed. The collisions are a bit fanciful, since the star is quite small in comparison to the space in the cluster. Nevertheless, the idea is for the star to get free of the cluster, as our own Sun did, and allow its planets to form and flourish—some of which may be Earth-like!

Home videos of the scripts in this volume could start by videoing the performance or the background for one or more scripts. Video clips or longer segments will be effective when projected as backdrop to the scripts, themselves. For example, a carefully timed and sped up sequence of the rising of the Guest Star in Script #1 can be made by putting together a set of images from the planetarium program, *Stellarium*, as mentioned in part 2(B).

Similarly, the explosions of the supernovae in Scripts #2 and #3 could be animated. There are examples of animated supernovae explosions on the internet, but students could personalize them. The hyper-velocity star making "a run for it" in Script #6 could also be shown emerging from a dramatic explosion near the supermassive black hole at the center of our galaxy. Script #11 calls for some creative animation as the structure of the two biosignature molecules are gradually revealed. The possibilities seem limitless—and fun.

Some students will enjoy writing not only computer code, but extended stories based on the scripts. For students in Creative Writing, a project might focus on writing short stories based on all the scripts. Finally, it would be possible to use each script in this volume as the kernel of a full-length novel. All of the characters in these scripts have family, and work, and adventures that can be imagined almost endlessly.

PART 3

SPACE SCIENCE AND ASTRONOMY SCRIPT PACKAGES

A. Supernovae Scripts

Without supernovae we wouldn't exist. The Big Bang at the beginning of the universe only made hydrogen and helium, and mere traces of the light elements, lithium and beryllium. These are the primordial elements. The oxygen, carbon, nitrogen, calcium, phosphorus, and others up to the element iron, in our bodies, were made at the centers of stars. The question is: How did they get out and into the molecular clouds from which our own Sun and its planets were formed? A further question concerns the heavier elements, like copper and zinc, which are present in small but very necessary amounts in our bodies. The answer to both questions is that they all formed and were dispersed through supernovae, like SN 1054 (Figure 1).

Figure 1. The A.D. 1054 supernova as seen from Earth, just before dawn on July 4. (Simulated using *Stellarium*.)

A supernova is the dramatic explosion of a star. It has so much energy that for a short while it can shine as brightly as its host galaxy. There are two main ways to produce this explosion. The first way requires two stars, one of which has gone through its life cycle and become a compact white dwarf star. The other star can be a less massive star that is taking longer to evolve, or a star that is merging with the white dwarf star. In each case, matter is transferred from the second to the first star, raising the core temperature of the first in a thermal runaway and triggering a massive explosion in what are called Type Ia supernovae. The incredible temperature and pressure of the explosion are enough to synthesize elements heavier than iron from the lighter elements.

The second way to produce a supernova is from the collapse of a star that is ten or more times as massive as our Sun. The star's nuclear fuel in its core has given out, and there is no more internal pressure to support it. It collapses dramatically, and its center reaches an incredible temperature. The whole thing explodes in a process that again synthesizes elements heavier than iron and nickel. These are called Type Ib, Type Ic, and Type II supernovae.

The explosion of a supernova drives most of its material out into the interstellar medium. This material is made up of both the elements produced by regular nuclear synthesis during a star's life and the elements produced in the instant of the supernova explosion. The material helps us in two ways. It contains all the elements needed on our planet and in our bodies. The high velocity material also pushes on the interstellar medium, that is, the gas and dust already present out beyond the star, so compressing it and making it readier to collapse and form a new generation of stars with their attendant planets.

The process goes on, with a new generation of massive stars giving rise to further supernovae and in so doing, building up and distributing the elements beyond the primordial hydrogen and helium. It has often been stated that, "We are all stardust." We are indeed made of material synthesized in successive generations of stars.

Remaining behind, at the original site of the supernova, is either a neutron star or a black hole. A neutron star consists of atoms in which their negative electrons under intense pressure have combined with their positive protons, so forming non-charged neutrons. It probably appears white to our eyes, because it radiates at multiple wavelengths. It is very compact, but it does not have quite enough mass to be a black hole. The latter is so compact that not even light can escape its intense gravity. We can't actually "see" a black hole with our instruments. We can only observe the effects of its concentrated mass on the trajectories of nearby stars or on any surrounding dust and gas. The gas and dust are accelerated so much just before they fall into the black hole that they give off tell-tale xrays.

Supernovae do not occur very often for us to see. The earliest recorded supernova was in A.D. 185. The Chinese also recorded them in A.D. 1006 and 1054. The remnant of the A.D. 1054 supernova has expanded out to form what we now call the "Crab Nebula." Europeans saw supernovae in A.D. 1572 and 1604. After that, telescopes made it possible to look for them in galaxies beyond our Milky Way. Supernovae that explode through the Type Ia mechanism are much sought after, because we think we have a good idea of how bright they actually are (their "absolute magnitude"), so they have become "standard candles." Since they are so bright, they can be seen a long way off, even to the limits of the visible universe with sufficiently large telescopes. In this way, they give us "cosmic distances." However, the theory of supernovae explosions needs to be refined. The time of the explosion is unpredictable, so to obtain data on a star just before, and at the very time of the explosion, is like "finding gold," as in our Script #2. Of course, gold is itself a product of supernovae!

The scripts can be presented in any order desired since they were written to stand alone. The authors have also demonstrated scripts by grouping them according to a theme other than the main headings (Supernovae, Milky Way, Stars, and Planets and Exoplanets), for example, Early Humans Scripts, #7 and #10. They have presented two at a time, or three, which, with discussion, provide about an hour of engagement.

Space Science and Astronomy Script Package #1
"The Appearance of a Guest Star in A.D. 1054"

Lessons in History and Culture

The supernova that created the Crab Nebula, called "SN 1054," was recorded by Chinese, Japanese, Korean, Arab, and some Native American astronomers. We rely on Chinese records because astronomy, which was often called "astrology" in ancient China, was most advanced in the Far East at the beginning of the second millennium. Nowhere else do the historical, archaeological, and artistic records of the A.D. 1054 stellar explosion provide quite such accurate and detailed data. Chinese astrologers—the astronomers of their day—left star charts and written records of SN 1054 that have been challenged repeatedly and not found wanting.

It is important to remember that documentation of "the Crab" took more than symbolic thinking. It required a type of explanation of natural events that characterizes human sentience and inspires all scientific, religious, and artistic thought (*cf* Rappaport and Corbally 2015). The interpretations of ancient Chinese scribes incorporate comparison, analysis, and creativity, and the result was the creation of "new cultural knowledge." In the script that dramatizes the sighting of SN 1054, Yang Weide, the Director of the Astronomical Bureau, explains to his assistant in the early morning of July 4, 1054, that he must be careful in his explanation of the new "Guest Star" for *political* reasons. He must flatter Emperor Renzong of Song, and be mindful of the Emperor's enemies. Indeed, according to Chinese records, Yang Weide did not officially report the Guest Star to the Emperor for two months, waiting until August 1054.

Lessons in Space Science and Astronomy

SN 1054 was probably a core-collapse, Type II supernova since it was visible in daylight for a couple of months before declining in brightness more slowly, only to become too faint to see at night with the eye by April 1056. Type I supernovae have a more rapid overall decline in brightness after their peak. These differences in light curve are the basis for distinguishing one type of supernova from another.

Historians are not in complete agreement about when SN1054 was first seen. Some European sources claim that SN 1054 was first seen in daylight, on April 11. The fact that it had declined to the brightness of Venus by July would fit the light curve of a Type II supernova. However, European sightings of SN 1054 still have many challengers, and none are widely accepted yet. We provide a scenario for a possible Script #14, in part 2(C), "How to Develop Additional Astronomy Script Ideas," in which an astronomer and an historian try to confirm or refute documentation of five European supernovae sightings that seem to have occurred around the time of the appearance of SN 1054. Their work does not take place until A.D. 2085, after new documentation is imagined to be uncovered.

While the Emperor was officially notified by the astronomer Yang Weide that the 1054 "guest star" appeared on July 4, an independent source from the history of the Kingdom of Liao marks the star's appearance at the time of a totally eclipsed Sun, on May 10. This would certainly have been an ill omen, since, according to Chinese cultural beliefs, the Song Emperor *was* the Sun. A later official sighting would have been much more propitious.

The remnant of SN 1054 can be seen today as the beautiful, filamentary Crab Nebula. In the core of this nebula is a pulsar, a rapidly rotating neutron star. This was the first pulsar to be identified with a historical supernova and the first observed pulsing in visible as well as radio wavelengths, at thirty times per second.

Worksheets for Use before a Script

I. Keywords

Astrological house	Peasant
Celestial	Prognostication
Edict	Prostrate; to prostrate oneself

"Guest Star"   "Standard candle"
Iridescent   Supernova
Nightingale   Temperate
Parasol

II. Questions for Investigation Before the Script

1. What is the "Crab Nebula"?
2. How are supernovae classified? How do they differ?
3. What is Chinese astrology, and is it the same as today's science of astronomy?
4. The Chinese capital at the time of this script was Kaifeng. Where is Kaifeng on a map of China today? What province does it lie in? What is the capital of China today?
5. The astrologer in this script—also called an astronomer—is Yang Weide. What is the family name in the name, Yang Weide?
6. Yang Weide shows great concern about making his Emperor content. Who was the Chinese Emperor in A.D. 1054?

Discussion Questions for Use After the Script

1. The action in this script does not take place in modern times. In what century does it take place, and how can you tell?
2. Had the telescope been invented by the time this action takes place? If yes, why doesn't Yang Weide use one? If no, how did people observe the stars?
3. Why was Yang Weide so worried about reporting the Guest Star to his Emperor?
4. Why would peasants, in general, be "restive" after a "rise in taxes"?

Costumes and Props for Inexpensive Productions

Costumes and props for scripts of exotic cultures can be quite easy to devise because it takes very little to suggest the nature of a culture to an audience. For the script on the A.D. 1054 supernova, a "blackout effect" can be achieved by the actors' wearing same-color pants and shirts (or turtleneck tops). The few costume pieces and props then stand out in contrast. For this script, which takes place in traditional China, in the year A.D. 1054, the male character, Yang Weide, is dressed in formal, aristocratic silk robes. However, a simple man's silk cap in a Chinese style and fabric (ours had a pigtail, too) is all that may be required to convey his social position and cultural identity. Men's Chinese caps are available from online vendors as party favors or Halloween costume pieces. Either character can wear a stand-up collar in the Asian style to convey Chinese identity. The female character can carry a paper parasol ("Chinese" or "Japanese" or "Asian" in online searches) as a prop. The parasol is important to this script, because it is written into the dialogue. Many inexpensive paper versions can be obtained from online vendors of wedding or baby shower decorations. One prop should not be used in this script: the telescope. In A.D. 1054, it had not yet been invented!

Introduction to the Script on the Sighting of the A.D. 1054 Supernova

The short script on the identification of the 1054 supernova is propelled by the court astronomer's tension about his impending astrological report to Emperor Renzong of Song. His assistant, Dong Zhi, is also concerned about her master, Yang Weide, fearing that he will not be sufficiently careful and thus put his life in danger. In retrospect, the historical Chinese records show that Yang Weide was extremely careful and clever to put a beneficent slant on his identification of a bright Guest Star, one whose appearance would normally cause great concern in the thinking of the times. Some of the words in this script are taken directly from Yang Weide's original report, which exists today in Chinese historical records. These words have been translated a number of times and the wording discussed thoroughly by translators and scholars. The words in the script are in some cases simplified versions of the original. The language in this script is often "courtly," conveying a sense that the two characters live and work in a feudal era, at the court of the Chinese emperor.

Script
"The Appearance of a Guest Star in A.D. 1054"

Characters:
YANG WEIDE, Chief of the Astronomical Bureau, during the reign of Emperor Renzong of Song.
DONG ZHI, who is his assistant.
EMPEROR RENZONG, who ruled 1022 to 1063. He is off-stage, but addressed by Yang Weide.

Place: Exterior.
Location: At court, in Kaifeng, China, outside the Astrologer's Observatory.
Time: In the latter part of the night, after the Guest Star appears, on 4 July 1054.

A [SLIDE OR SIGN] gives the year, place, and time:

"A.D. 1054, July 4. Kaifeng, China. 3:30 a.m. in the morning"

YANG WEIDE

Foolish woman! Why do you carry a parasol when there is no sun?

DONG ZHI

Oh, celebrated astrologer, I awoke to the sound of a nightingale who sang too loudly for this early hour of the morning. And lo! I see the cause of the bird's misbehavior! Forgive me for interrupting your celestial observations...

YANG WEIDE

Yes, the light! Look at it! See how bright it is! It is surely an omen, but of what, I do not yet know...

DONG ZHI

You are troubled, Yang Weide? Surely a star that shines so brightly can mean nothing but good fortune!

YANG WEIDE

I must think carefully, woman, and I must act even more carefully! The Emperor's enemies gather to the north, and his bribes to leave us in peace have not exactly worked...

DONG ZHI

I know the peasants are restive after a rise in taxes—

YANG WEIDE

—It's true.

DONG ZHI

But, what can you do, oh noble Astrologer, gifted with your special knowledge of the stars and able to foretell the future with your prognostications? How can *you* ease the Emperor's burden?

YANG WEIDE

There are ways to handle such a man—a man whose arm is strong but whose enemies are stronger.

DONG ZHI

Oh wise Astrologer, be temperate in the advice you give. Do not make the nightingale's mistake and sing out at the wrong time.

YANG WEIDE

The dawn comes, Dong Zhi. See how the Guest Star lingers in the sky for all to see, even in the day!

DONG ZHI

I will take my leave of you now and return to the palace. But do be careful, Yang Weide!

Dong Zhi bows and exits.

Yang Weide turns away from the audience. A new [SLIDE OR SIGN] replaces the old one, reading:

"A.D. 1054, August 27. 10:30 a.m. in the morning."

YANG WEIDE

He is speaking to the Emperor off-stage.

Prostrating myself, I report that I have observed the appearance of a Guest Star! I respectfully submit that the *Prognostications in Respect of the Emperor* read that the Guest Star had a slightly iridescent yellow color.

Respectfully, I have prognosticated, and the result said: The Guest Star does not infringe upon the wrong astrological house. This shows that a Plentiful One is Lord. If the guest star does not trespass, an Abundantly Enlightened One is in office!

I request that this prognostication be given to the Bureau of Historiography.

He bows his head and stays in place.

DONG ZHI

She steps in without a parasol, and with hands in prayer position, announces the following.

And so, all the officials presented their congratulations to Yang Weide, and by Imperial Edict, it was ordered that the prognostication should be sent to the Bureau of Historiography.

The Guest Star stayed with us for two years.

Space Science and Astronomy Script Package #2
"How Serendipity Leads to Theory: The
A.D. 2054 Eta Carinae Supernova"

Lessons in History and Culture

Our second dramatization of the sighting of a supernova takes place one thousand years after the A.D. 1054 supernova, in A.D. 2054. Like a number of scripts in this book, the story is fiction, but not the fanciful kind, and is instead based on projections of today's events. For example, the script takes place at the European Southern Observatory, which exists today. However, the action takes place at the fancifully named, Nquillatún, the *fifth* unit of the ESO's Very Large Telescope. In truth, there is no fifth telescope today—there are only four. However, the name of the projected fifth telescope follows the custom of naming the ESO's telescopes after spirits in Mapuche Indian mythology. Another telescope is also mentioned in the script and it, too, has a Mapuche Indian spirit name. Antü is a male spirit and the most powerful Pillan. The Pillan are good spirits, but they also punish with drought, earthquakes, and disease. Antü is an existing telescope at the ESO facility. The name is sometimes interpreted as meaning, "the Sun."

The European Southern Observatory on the Atacama Desert of highland Chile is a cold, bleak, dry place, but it is perfect for astronomers because of the absence of light and other pollution. It affords a spectacular view of the Milky Way in the Southern Hemisphere. The characters in the script are both astronomers, one named Henry and another named Xiao-Xing. Both of them have Chinese heritage, so they know well the value of delicious *dim sum* (small meat and vegetable pastries), and they offer a gentle bribe of *dim sum* to their colleagues, to gain time on all five telescopes. In this remote part of the Andes Mountains, *dim sum* is a treat.

The script also harkens to ancient, but still extant, Chinese culture in the discussion by these two astronomers of astrology, revealing the continued fondness of some Chinese expatriates (and many others) for astrology in modern times.

The result of all these cultural factors is a blend of cultures that we see so often among astronomers who often travel great distances to obtain "clear skies" and so, the best data possible from widely dispersed large telescopes. They can spend months and years away from their native countries for the sake of their research projects. They carry their native cultures with them, in part, and they learn local customs from the people who inhabit the places where they work.

Lessons in Space Science and Astronomy

It is impossible to see into the nebula, Eta Carinae, but astronomers have proved it contains two massive stars (Figure 2). The binary system is highly unstable. One star is about thirty times the size of our Sun, and the other, around one hundred and twenty times the mass of the Sun. Stars tend to "go supernova" at more than ten masses of our Sun, so both stars are likely to explode soon—the large one, sooner than the smaller. In the following script, this is what we envision.

The system underwent some "almost supernova" eruptions over the years around A.D. 1841. A still earlier brightening was seen in the mid-eighteenth century. Eta Carinae continues to brighten and fade on a time scale of five and a half years, which is the revolution period of the two stars round each other, although the magnitude swing is less than in the two previous eruptions. The Eta Carinae nebula should be watched regularly, both for its beauty and for its potential to enhance the science of the most massive stars, and of supernova explosions.

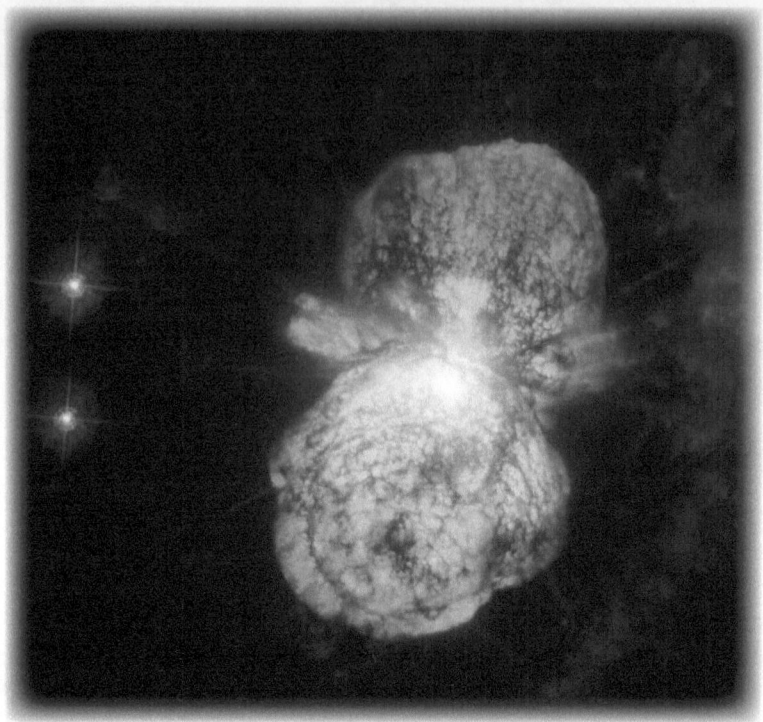

Figure 2. The Homunculus Nebulae surrounding Eta Carinae.
(Credit: ESA/NASA)

Worksheets for Use before a Script

I. Keywords

Actuary	Light pollution
Astrology; astrological chart	Nebula
Dim sum	Scientific prediction
Empirical data	Serendipity
Fortune telling	Supernova (Supernovae)
Homunculus	Theory
Hypothesis	Unstable star system

II. Questions for Investigation Before the Script

1. Do today's scientists understand why stars explode? Please explain.
2. What is unusual about the nebula, Eta Carinae?
3. Why is the Atacama Desert such a good place for astronomers to work?
4. What does the "control room" for a modern telescope have in it?
5. Which big science prizes are given in Stockholm, Sweden?

Discussion Questions for Use After the Script

1. The Chinese name "Xiao-Xing" means "morning star." In the script, the character Xiao-Xing says, "If you substitute average values for all the variables you *can't* measure in Eta Car... then, you've got one of the most unstable star systems we know." How is her statement an example of scientific thinking? What is she developing in her thoughts and words?
2. Why does Xiao-Xing conclude that Henry "is famous" when he tells her about the data he captured on the exploding star *just before* it explodes?

Costumes and Props for Inexpensive Productions

When the authors performed the Eta Carinae script, they used informal shirts and pants. This is often the dress of astronomers working in control rooms, which are now thankfully well heated! However, the notion of cold surroundings on the high Atacama Desert can be suggested by the two characters' wearing knitted ski caps with ear flaps, which, in shape and design, are much like the caps of the local Indian populations. Xiao-Xing enters wearing one of these caps, stamping her feet from the cold. Henry has retained his cap on his head. Knitted ski caps are available for modest cost from many online vendors.

Introduction to the Script on a Possible Eta Carinae Supernova in A.D. 2054

The source of tension in the script lies in the possibility—by no means a high probability—that one of the five stars that Henry is watching (with the five ESO telescopes) will "go supernova" while they are observing them. The characters' elation over the fact that one does explode, and that Henry achieves what no astronomer had yet done—capturing the data on a star just before it explodes, as well as while it explodes—is the true source of drama and excitement in this script. In the end, he wins the Nobel Prize for his work, and his colleague Xiao-Xing foretells this event in her exclamation, "Oh, Henry!" when she first realizes that one of the stars has exploded and he's been watching it all the while.

A backdrop showing a modern telescope's control room on a projected slide is an easy way to illustrate the surroundings for the characters in this script. The amazing explosion of Eta Carinae could show up as successive images that are either direct or spectroscopic. In the performance by the authors, they showed an animation of an exploding star. If spectroscopic data were imaged, then screens would contain pictures of a changing spectrum, or a plot of spectroscopic output.

Script

*"How Serendipity Leads to Theory: The
A.D. 2054 Eta Carinae Supernova"*

Characters:
HENRY YANG, Astronomer
XIAO-XING ("Morning Star") MING, Astronomer

Place: Interior, Control Room, Nquillatún, fifth unit of the
    Very Large Telescope, operated by the European Southern
    Observatory.
Location: Cerro Paranal, Atacama Desert, northern Chile.
Time: In the latter part of the night, 10 July 2054.

---

*A [SLIDE OR SIGN] gives the year, time,
and place:*

"A.D. 2054, July 10. 3:30 a.m. Nquillatún
Unit Telescope, Atacama Desert, Chile"

XIAO-XING

*She rushes in, shoulders bent against the
cold.
She stamps her feet and rubs her hands.*

Brrr! It's cold out there! Not as cold as Mars
will be, but the Atacama Desert is *cold!*

HENRY

He swivels from his control panel.

Xiao-Xing! The Morning Star comes early tonight!

XIAO-XING

Silliness! You know very well I have no control over what my parents named me! You seem to think that the name "Xiao-Xing" is some kind of *prognostication* that I would be an astronomer! ...I mean, how can it be? ...You don't believe in fortune-telling, do you, Henry?

HENRY

I don't know, Morning Star. You can take the boy out of China, but you can't—

XIAO-XING

—You can't take China out of the boy! Yes, I know...So, tell me, *do* you believe in astrology and telling the future by the stars? You don't, Henry, do you?

HENRY

He smiles.

Well...What's the harm in it? ...I *do* believe in scientific prediction based on data of past events: Not always perfect, but it's good. So I'm not so sure, Morning Star, about the answer to your question.

XIAO-XING

Henry...tell me, do you consult your astrological chart every day?

HENRY

No, not *every* day.

XIAO-XING

Ah-ha! But you do consult your astrological chart... *sometimes!* Right?

HENRY

He shrugs.

My mother is an actuary in Shanghai. No one relies on data like actuaries! And *she* still visits *her* astrologer.

XIAO-XING

Henry, how can you be a scientist—how can you *gather* data and *analyze* data— and still believe in astrology? You're *a scientist!*

HENRY

The belief systems are very different, but equally old, Xiao-Xing.

XIAO-XING

But, Henry, one is based on carefully collected, *empirical* data and the other is based on...on... fantasy! Magic!

HENRY

He turns back to the control panel.

You're right, of course. ...But, do me a favor, Xiao-Xing. I want you to look at my short list. *You* know stars, Morning Star!

XIAO-XING

Henry...You're avoiding the subject!

HENRY

He ignores her protests.

You see, I'm here at NquillatÚn, the fifth unit of the Very Large Telescope, and I've got a list of *five unstable star systems* that just might—according to everything we know now—blow! So, I needed 5 telescopes

and I had to bribe the operators of the other
4 unit telescopes to give me a few hours—

XIAO-XING

—*Bribe?* Henry Yang!

HENRY

With *dim sum.*

XIAO-XING

She stares at him, then bursts out
laughing.

Dim sum?

HENRY

Freeze dried, straight from my mother in
Shanghai.

XIAO-XING

You're kidding...

HENRY

It's hard to get *dim sum* on the Atacama
Desert, Xiao-Xing.

XIAO-XING

Brilliant.

HENRY

He gestures to one of his screens.

Anyway, over here at the telescope called *Antü*—

XIAO-XING

—These Mapuche Indian names always confuse me, Henry...

HENRY

At Antü, I've got the instrumentation trained on Eta Carinae, and—

XIAO-XING

Eta Car's a strange one, alright!

HENRY

Strange enough. According to what we know, it's unstable, but we still can't see inside that nebula, the white one shaped like a peanut—

XIAO-XING

—The one shaped like *a little man!*

HENRY

Yes, they call it an *homunculus*, like a bent-over little man.

XIAO-XING

Odd.

HENRY

Odd, and unstable—*I* think! Don't *you?*

XIAO-XING

Hmm… I remember answering this question in my PhD exams. If you substitute average values for all the variables you *can't* measure in Eta Car, because of the nebula shaped like a little man, then you've got one of the most unstable star systems we know!

HENRY

Theoretically, it would take only one or two of those variables to be *out-range—not* average—and you might have a supernova! Yes?

XIAO-XING

Now you're starting to sound like a scientist! There's no need for astrology tonight!

...But... to answer your question...

She stops, stares, and points, mouth open.

Henry—

HENRY

—What? What's the matter?

XIAO-XING

Henry, look at the screen from Antü...

HENRY

What?

He swivels.

Oh! It blew! Xiao-xing, look at that! I've got to get the other four units trained on Eta Car... Quickly, help me here!

XIAO-XING

She rushes to the instrument panel to help him.

I'm on it! ...The stars were aligned tonight, Henry!

HENRY

I thought you didn't believe in astrology, Xiao-Xing. But, you *are* right. The stars were aligned to produce good fortune tonight! And, you know something else?

XIAO-XING

What?

HENRY

We've got data from that star system for four hours *before it blew!*

XIAO-XING

She stops, and holds her hand to her forehead.

Oh, Henry! ...You're famous!

She lowers her chin and walks off stage.

HENRY

*Henry turns to address the audience
alone, and with a serious tone,
announces the following.*

Great fortune visited us that night. The stars were aligned correctly. But it was many years before I analyzed all of the data from Eta Carinae and built a theoretical model of how and why star systems explode... go supernova... so to speak. You see, the clue was found in the data I captured before the star system exploded. No one had ever been able to capture that before. And so, eventually, I analyzed the data, but then, it was ten *more* years before I received the *big prize.* ...My parents were very proud. They came all the way to Stockholm for the ceremony.

He bows from the waist, and exits.

Space Science and Astronomy Script Package #3
"The Cassiopeian Supernova of A.D. 3054,
or, Good News Travels Fast"

<u>Lessons in History and Culture</u>

The third and final script about the sighting of a supernova looks far to the future and imagines, through dramatic means, the explosion of Rho Cassiopeiae in A.D. 3054, two thousand years after the appearance of SN 1054. The dialogue relies on a form of communication that does not yet exist: a sub-space channel that allows always-chatty human beings to communicate at a speed faster than light. An anthropologist doing research on Jupiter's moon, Europa, contacts her friend, an astronomer on Earth's Moon. She tells him about a new supernova and she advises him to train his telescope "in the direction of Rho Cas," a star in the constellation Cassiopeia. Both characters know that Earth's Moon will receive the light from the explosion of Rho Cas forty minutes after they pick it up on Jupiter.

Figure 3. "Rho Cas" going supernova, as seen from Jupiter's Moon Europa, A.D. 3054. (Simulated using *Stellarium*)

It is a "golden opportunity" to observe the supernova when the luminous peak is strongest but most short lived and in the ultraviolet light range (only possible beyond the Earth's atmosphere). The astronomer in the script changes his instrumentation just in time to catch "the shock breakout through its massive shell." He notes that Rho Cas, well known as a highly variable supergiant star, had been projected to explode "quite soon," according to the science of A.D. 3054.

At both our present time and at this distant time in the future, the main mystery concerning supernovae like SN 3054 (which is expected to be a Type II-P supernova, like SN 1054), is this: What exactly happens when their core collapses and they "go supernova"? Observing the shock breakout, which is proportional to the mass of the progenitor star, would help decide between the competing scientific scenarios on core collapse and the fate of the subsequent ejecta.

It is noteworthy that the characters incorporate the new stellar events into their scientific frameworks easily. The capacity for humans to do so is what makes them so special. Cultural systems, like the symbolic frameworks used to express them, are infinitely expandable and capable of incorporating new elements very rapidly when it is necessary to do so. The script illustrates the use of existing belief systems, over a thousand years in the future. And yet, this is but a short time in comparison to the long history of so-called "modern thinking" or "symbolic thinking" for the species *Homo sapiens*, who probably emerged in East Africa, with something like these abilities, around 200,000 years ago (McDougall, Brown & Fleagle 2005).

The two friends in the script appreciate the startling size and brightness of the explosion in the constellation of Cassiopeia. Together, they realize how important it is to capture as much data as possible, as early as possible. The astronomer—using science that does not yet exist—notes that astronomers were "expecting" the star to explode at any moment. One can only imagine the art forms and spiritual commentary that would go on to incorporate the Rho Cas supernova, in A.D. 3054, and thereafter. Rho Cas explodes within a framework of greater human understanding of the science involved in supernovae. Still, they are fascinated with the new bright light in the black sky and they cannot help but call it "beautiful."

Lessons in Space Science and Astronomy

Rho Cassiopeiae is an example of a highly unstable star. It is not surrounded by a nebula, like the pair of stars in Eta Carinae, so we can see its spectrum and determine that it is a yellow hypergiant star. It has about the same surface temperature as our Sun, but its diameter is around four hundred fifty times our solar diameter and its luminosity is a huge, half-million times solar. It is not quite as unstable as the Eta Car pair, but it periodically has comparable eruptions of material from its surface, which change its brightness by a factor of around four. Rho Cas is a star with a mass between fifteen and thirty times that of our Sun, and it is quite near the end of its evolution. It will certainly have a core collapse and go on to produce a supernova Type II, but we may have to wait a bit, say, a thousand years!

Supernova remnants (SNRs) such as the Crab Nebula are called "plerions," for which the exploding material more or less evenly fills an expanding bubble around the original star, now a pulsar. Another kind of remnant is one for which the material is more concentrated in a shell at the edge of the bubble and there is no pulsar at the center. A third kind has characteristics of both a plerion and a shell, and is known as a "composite." Current thinking by astronomers is that the SNR of a Type Ia supernova will be a shell and the SNR of a Type II will be a plerion. Therefore, we guess in the script that Rho Cas will produce a plerion. By A.D. 3054, the theory may be much clearer on this, and the characters won't have to guess.

The positions of the Earth and its Moon, Jupiter and its moon, Europa, and the exploding star, Rho Cas, on 10 July, A.D. 3054, are important to the script's dramatic tension. Jupiter will lie almost exactly along a line between Earth and Rho Cas, and, about forty light-minutes away from the Earth. That means that the explosion of Rho Cas as a supernova will be seen on Europa about forty minutes before the same light reaches the Moon. Any warning of the event obviously has to outpace the speed of light, and therefore the need for "sub-space communications" in the script.

Worksheets for Use before a Script

I. Keywords

Anthropologist	Proclivity
Europa	Spectrometer
Jovian	Shell (supernovae)
Kelvin (temperature)	"Sub-space communications"
Plerion	Vatican Observatory

II. Questions for Investigation Before the Script

1. Do Earth moonbases exist now? Do Jupiter moonbases exist now? Why, or why not?
2. Will Earth moonbases exist in the future? Will Jupiter moonbases exist in the future? Why, or why not?
3. Can human beings communicate at a speed faster than light today? Will human beings be able to communicate at a speed faster than light some day? Why, or why not?

Discussion Questions for Use After the Script

1. In what century does the action in this script take place?
2. What was Suzanne doing out on one of Jupiter's moons? What was the advantage of her using "sub-space communications" to contact her friend, Peter?
3. Sketch out the geometry of the Eta Car supernova, the Jovian moon Europa, and both Earth and Earth's Moon.

Costumes and Props for Inexpensive Productions

When performing the script on the Rho Cas supernova, the authors wore contrasting clothes: One in all-black and the other in all-white. They then donned "space goggles" of matching black and white, which were obtained online by searching for "space costumes." The goggles were very sleek, fit the script, and were inexpensive. The only problem was that the sunglasses (what they really were) were too dark to read the scripts! Therefore, after a laugh from the audience when they saw the all-black and all-white outfits, the "space goggles" were removed without a hitch. If students are memorizing lines, instead of reading them, that problem would not arise. The variety of space costume pieces online was surprising, and while some were expensive, most were not. An image of the authors wearing "space goggles" is at http://thehumansentienceproject.org.

The authors performed the script back to back, to give an impression that the two characters were a great distance apart, and were only experiencing one another over a communications channel.

Slides with the landscape of the frozen moon Europa, and the explosion of Rho Cas (with a calculated brightness suitable for our Sun's position) were good accompaniments (Figure 3).

Introduction to a Script on a Possible Cassiopeian Supernova in A.D. 3054

This script introduces a very long-distance conversation between two friends: a female anthropologist and a male astronomer. However, the sex of the participants can be switched, as for most of the scripts in this volume. The way it is presented here, the female anthropologist is researching the effects of "sub-space communications" (communicating at faster than light speed) on humans and their culture on Europa. She calls an astronomer who is in residence at "Vatican Observatory 3," which does not exist today. There are only two installations of the Vatican Observatory, one in Castel Gandolfo, Italy, and the other, in Tucson, Arizona. If desired, the observatory's identity can be changed to an observatory that is local or closer to the students participating in the script. However, students should speculate on changes in that locale by the year A.D. 3054, and given reasons for their projections as part of the discussion. It is a good exercise in futures research.

The source of tension in this script is the time pressure that both friends are under—the anthropologist on Europa, who knows she must warn her friend soon of the coming light from the explosion of Rho Cas, and also the pressure on the astronomer on Earth's Moon to switch his instrumentation rapidly toward the star, in order to catch the explosion. The final reveal involves the astronomer's reaction of delight and wonder when he "catches the light from the explosion," and the anthropologist's equal delight in the beauty of the supernova.

Script
"The Cassiopeian Supernova of A.D.3054, Or, Good News Travels Fast"

Characters:
PETER CORBALLY, SJ, Director of the Astronomical Bureau, Vatican Observatory Three.
SUZANNE RAPPAPORT, Contract Anthropologist, Europa Lunar Government.

> Place for Rappaport: Interior, Habitation Module 12, Europan Jovian Colony.
> Place for Corbally: Interior, Vatican Observatory Three, North American Moonbase, Earth's Moon.
> Location: At their respective sub-space communications stations.
> Time: In the latter part of the night, Greenwich Mean Time, 4 July 3054.

A [SLIDE OR SIGN] gives the year, time, and place:

"A.D. 3054, July 4. 3:30 a.m. GMT. Europan Colony, Jupiter, and Vatican Observatory Three, Earth's Moon"

PETER

Hi, Suzanne. You called to give me a weather report, didn't you?

SUZANNE

You're closer to Sol that I am! You should be giving the weather report to me! But I'll humor you... It's 110 degrees outside...

PETER

That's Kelvin, isn't it? My goodness! That's *cold.*

SUZANNE

She laughs.

Indeed, Peter! In any case, the storm on Jupiter's surface has calmed down a bit, so we've got clear communications. You sound like you're right next door, Peter.

PETER

I still have a hard time believing in sub-space communications. It's just not normal for human beings to communicate at faster than light speed! I prefer to think we're just reading each other's minds.

SUZANNE

But Peter, you're the celebrated Astronomer! You understand these things better than I do. I just know what Sub-Comm has done to humanity's proclivity for chatter, in this noticeably smaller solar system... That's why this Anthropologist is out here on this hunk of ice!

PETER

It's still creepy... That's a technical term, of course. Besides, I want to know how your work is going. Will you come to the Moon before returning to Earth?

SUZANNE

Maybe... But... There's something important going on and I don't have much time to tell you! Are you at the observatory? Do you have that telescope humming? If so, train that baby in the direction of Cassiopeia!

PETER

Hmm. Your voice changed. This is important, isn't it?

SUZANNE

It's *beautiful!* Take a look at Rho Cas, Peter! Or what *used to be* Rho Cas!

PETER

You're kidding... You're not! You're NOT kidding! Are you sure? We've been expecting that one to blow for centuries! All the models said it would...

SUZANNE

We think you've got a likely plerion coming your way, Peter. Get ready for it! The light should reach you in about... two minutes.

PETER

You've had it for 35 minutes! Are your guys catching the data? Hold on! I don't have much time... I just need to get the UV... fast-frame spectrometer going...

SUZANNE

...Keep jabbering, Peter. This line won't stay open without human conversation...

PETER

...to get the details of... the shock breakout through its massive shell. ...No robot-speak, Suzanne?

SUZANNE

Only human-speak, Peter. It's designed that way.

PETER

Here's the light pulse! I've got it! Oh! Oh, it's bright!

SUZANNE

Forgot to tell you that.

PETER

It must be near -8! I think it's brighter than 2054! It'll be waking up the folks down on Earth any second now! Wow! I've got to

go, I've got to get the data streaming. I'll
speak with you—When?

SUZANNE

When I get there. I'll be back through the
North American Moonbase in two months.

PETER

Wow. Okay... Let me go. See you then.

SUZANNE

See you under a new star, Peter.

PETER

A Guest Star, isn't that what Yang Weide
called it, back in ancient China? What a
guest! This one isn't going to smell after
three days, is it?

SUZANNE

Signing off. See you soon.

PETER

Signing off. And thanks!

B. Milky Way Scripts

The Milky Way is our own, spiral galaxy. Just as the solar system is home to our Sun and its planets, so the Milky Way is home to the around two hundred billion stars with which we travel through the universe. An important component of our Milky Way becomes clear when we look at a similar galaxy edge-on, like the NGC 3982 Galaxy (Figure 4). Interweaved with the bright "star-forming" patches of the latter are dark lanes, where dust obscures the light of young, embedded stars. That dust is accompanied by gas, and these two are often in molecular form, so the locations are called "molecular clouds."

Figure 4. The NGC 3982 Galaxy, a Spiral Galaxy Like the Milky Way. (Credit: NASA/ESA)

The shape of the Milky Way looks like two soup plates, one inverted and placed on top of the other. When seen from the side, there is a bulge in the middle of the disk. Our solar system is located a bit more than halfway out from the center of the Milky Way, at about twenty-seven thousand light years. That makes our galaxy's diameter around one hundred thousand light years. If you could look at the Milky Way from the top, it would show a short bar of stars in the center, and then spiral arms containing stars and molecular clouds would spread out from each end of the bar. NGC 3982, shown in Figure 4, does not seem to have a central bar, and it is about one-third the size of our galaxy. The entire Milky Way is rotating, and at the position of the Sun, it goes around once every quarter of a billion years. This gives us a velocity of two hundred and twenty kilometers per second. So hang on tight!

The Milky Way is surrounded by a less dense spherical halo of stars. Some of the halo stars were formed before our galaxy collapsed down into a disk, when it looked more like a sphere. These primitive stars are mostly in tight clusters, called "globular clusters," containing up to one hundred thousand stars, each. In the halo, there are also filaments of stars, whose characteristics indicate they have been captured from smaller, dwarf galaxies. Looser groupings of stars—the "open clusters"—are found in the disk of the Milky Way, and these are sites where stars have been most recently formed.

To locate any object within the galaxy, we use a "galactic" two-coordinate system, centered on the Sun and with its prime axis towards the center of the Milky Way. This is similar to using "bearing and elevation" when aiming a large gun. The other necessary location information is the object's distance from us. Some stars change their positions very rapidly and are called "high-velocity stars."

In the very center of the bulge of the Milky Way there is a supermassive black hole. It cannot be seen directly, but only deduced from the rapid rate of motion of stars nearby or, in the case of NGC 3982, from the high-energy radiation it emits.

Beyond our Milky Way lie other galaxies in the visible universe. There are perhaps ten times as many of them as there are stars in our own galaxy, one to two trillion galaxies. Like stars, galaxies tend to cluster in various sized groupings, coming under the influence of their mutual, giant gravities. The Milky Way belongs to the Local Group of around fifty-four galaxies. Most are dwarf galaxies, with the Milky Way being the second largest in our Local Group, after the Andromeda galaxy.

Space Science and Astronomy Script Package #4
"Giant Molecular Clouds Mirror Ancient Inca Animals, A.D. 1463"

<u>Lessons in History and Culture</u>

A familiar image of Inca civilization is the site of Machu Picchu, whose purpose is still a mystery. The center of the Inca Empire was the location of present-day Cuzco, Peru. Machu Picchu, the locale for the first script, may have been a summer palace or it may have simply been one of a string of stopping places that tied together the vast Inca Empire.

The Inca people emerged in the Cuzco Valley around A.D. 1000. Their empire consolidated gradually and reached its maximum just before the entry of the Spanish conquistadors in the 1500s. The Inca conquered one after another of their neighboring tribes until a period of rapid expansion under three Inca emperors between A.D. 1438 and 1527. The empire stretched over 3,400 miles down the length of the Andes Mountains. At the height of the empire, the Inca people numbered about 100,000, but they ruled 10 to 12 million people from at least eighty-six ethnic groups. Their power all came to a sudden end at the hands of a small band of Spanish soldiers in A.D. 1532. Anthropologists and historians have learned the most about the Inca from the diaries of the conquistadores and the Inca graves that were discovered after the collapse of the empire. Much of the fine Inca craftsmanship was destroyed by their Spanish conquerors, who had guns, horses, and writing.

Pachacuti was the ninth Sapa Inca (or "emperor"), and he lived from A.D. 1438 to 1471. He, along with his wife, Mama Anawarkhi, are characters in the first script. He, his father, and his son brought diverse Native American tribes under a single political umbrella. One of the mechanisms for consolidating the empire was to impose the worship of Inti, the Sun God, on the conquered tribes. At the same time, local people also had a tendency to keep their traditional gods, too. You'll hear about this in the script, as well as something about the political task of keeping a far-flung empire together in a time of foot transportation and no writing.

Lessons in Space Science and Astronomy

It was not until the 1920s that observers established that many of the nebulae seen in the night sky were in fact "island universes," a term used by Immanuel Kant. They are now called "galaxies," and they contain billions of stars with accompanying gas, dust, and stellar remnants, plus dark matter that we cannot see but whose gravitational effects on other galaxy components we can detect.

The Milky Way is the galaxy containing our own Sun and planetary system, and its name comes from the milky band across the night sky that can be seen when away from city lights. Patches of the Milky Way appear dark and starless, and these are the locations of giant molecular clouds of gas and dust, which absorb the light of the stars behind them. They are not to be confused with dark matter, whose composition we do not know.

It is usually the patterns of bright stars that humans make into constellations. However, in a really dark sky, such as the sky available to the ancient Incas, the dark patches in the Milky Way will appear so striking that they suggest animal shapes. It is like imagining shapes in fluffy clouds against a background of the blue sky, but in reverse.

The center of our galaxy is near the constellation of Sagittarius, which goes directly overhead when we are at latitude -30 degrees, around 20 degrees south of Machu Picchu. As expected, the center of the Milky Way across the night sky is its brightest part, despite all the dark patches from the molecular clouds. The Milky Way is especially impressive in the highlands of Peru and Chile, compared with the view from the Northern Hemisphere, so the ancient Incas (like the people living in the Andes today) were treated to a spectacular night show.

Worksheets for Student Use before a Script

I. Keywords

Artisan	Irrigation
(Social) Class	Llama
Conquest	Milky Way
Empire	Molecular clouds
Galaxy	Planetary science
Goldsmith	Rainy season

II. Questions for Investigation Before the Script

1. Where was the Inca Empire? In what years did it exist?
2. Which present-day countries did the Inca Empire cover (in whole or in part)?
3. How big was the Inca Empire at its maximum?
4. How do the views of the Milky Way differ in the Northern and Southern Hemispheres?

Discussion Questions for Use After the Script

1. Where were the "llamas" that Pachacuti describes? What are they made of?
2. Is the Milky Way a planet, a star, or a galaxy?
3. What's the position of our Sun in the Milky Way?
4. Why would the Earth's Moon remind Anawarkhi of things that "come and go"?
5. Why does the Earth's Moon change shape? On what cycle does it change shape? Why?
6. What is the "Man in the Moon," in terms of planetary science?

Costumes and Props for Inexpensive Productions

Simple costumes and props can give a good sense of the Inca people, especially the nobility. We know that the Inca in the higher classes sometimes wore headdresses with feathers. We also know that emperors wore a great deal of gold and silver ornamentation. "Gold" necklaces that look like Inca breastplates can be acquired at online vendors for very little cost, and feather headpieces and masks are available from companies who sell Mardi Gras paraphernalia. Simple tunics can be fashioned from rectangular pieces of inexpensive material that are simply folded over and tacked at the sides. A hole is cut for the head.

Introduction to the Script on the Cosmology and Politics of the Inca in A.D. 1463

The following script has two characters: Pachacuti, the Inca emperor, and his wife, Mama Anawarkhi. The action takes place on the evening of December 3, in the summertime of the southern hemisphere. The two characters are at their palace at Machu Picchu, which was then thriving and home to many people.

The source of tension in this script lies in the burdens of leadership and Pachacuti's concern that he and Anawarkhi serve as role models for the people they rule, by honoring Inti, the Sun God. She continues to honor the Moon Goddess, too, who is named Mama Quilla, and her husband goes along with this, to a point. In the end, tension is heightened by a rebellion of silver miners from a conquered territory, and we see that Mama Anawarkhi is quite politically astute in her own right.

The script plays well with a backdrop of the night sky from the southern hemisphere. Animal images of two llamas, a bird, a snake, and a toad can be picked out from patterns of molecular gas in the Milky Way, which show up as dark portions. The characters discuss how important these animal spirits are to them, showing that astronomy and culture are interwoven in the characters' understanding of their cosmos.

Script
"Giant Molecular Clouds Mirror Ancient Inca Animals"

Characters:
PACHACUTI, Ninth Sapa Inca of the Inca Empire, who lived A.D. 1438 to A.D. 1471
MAMA ANAWARKHI, his Wife

> Place: Interior, Living Quarters of the palace, Machu Picchu, northwest of Cusco.
> Location: Andes Mountains of present-day Peru, in South America.
> Time: Evening, 3 December A.D. 1463.

A [SLIDE OR SIGN] gives the year, time, and place:

"A.D. 1463, December 3. 9:30 p.m. Machu Picchu, palace of the Sapa Inca, Pachacuti, and his wife."

PACHACUTI

His arms are folded, raised in front of him, in a "military" stance.

Anawarkhi!

He stamps his foot.

Maaa-maaa A-na-warrrk-hi!

MAMA ANAWARKHI

She rushes in, bowing.

Yes, my husband, do not be angry with me. I was outside speaking with Mama Quilla!

PACHACUTI

Goddess of the Moon...

MAMA ANAWARKHI

Yes, my husband.

PACHACUTI

Anawarkhi, you must not let the priests and priestesses see you! We have just expanded Coricancha, the Temple of the Sun in Cusco. We must set an example for the priestly and noble classes, and honor Inti, the Sun God! They will worship Inti above all the other gods! The unity of the empire depends upon it!

MAMA ANAWARKHI

Yes, my husband, oh wise and fearless conqueror, brave warrior who defeated the Chancas and the foreign tribes to our

North and South, East and West! You have forged a great empire, Pachacuti!

She comes out of character momentarily and addresses the audience coyly.

That Pachacuti—he's such a cutie...

She resumes her character.

But, husband, the rainy season begins soon and the great festivals commence. I cannot forget my old spirits, like Mama Quilla, goddess of the Moon...

PACHACUTI

I understand, Anawarkhi. You have your own kind of wisdom, just like the foreign tribes—and all the workers and artisans they provide to us! We do not insist that they disavow their old spirits, only that they worship Inti above the others!

MAMA ANAWARKHI

She walks to the veranda and sweeps her arm upward to the spectacle of the Southern Hemisphere's Milky Way [on SCREEN].

Look at the river! Look at the river of stars, husband. The gods that live there must be honored so that we have a plentiful harvest. You know this, yourself, Pachacuti. Tell me you do not.

PACHACUTI

He has followed her and gazes upward with her.

When I was a boy, my nana taught me about the mother llama and the young llama right beside her.

The llama and baby llama are illuminated [on the SCREEN.]

The figures appeared just recently, and I knew then that the rains would start soon, and with the rains, water for our irrigation, so we can grow enough to eat—

MAMA ANAWARKHI

—See the toad! He makes me think of the rains, too! And he always makes me laugh...

PACHACUTI

You know, Anawarkhi, I still look up and when I see the llamas, I feel reassured. It's strange...

MAMA ANAWARKHI

These are not strange thoughts, husband. They are natural. It's the same with Mama Quilla. She reminds me that things come and go, and when they go, they come back again, just like the Full Moon.

PACHACUTI

He speaks in a confidential tone.

I have made sure the artisans include the llamas in their sculptures for Coricancha, even the golden ones... But, they are small and hidden. The goldsmiths mold the llamas into the finest pieces we have. The llamas are still there, but Inti is the biggest and the grandest!

MAMA ANAWARKHI

She laughs and looks at him askance.

I wonder, Pachacuti, if we should tell everyone that you do this...

A terrible din of noise is heard off stage, shouting and commotion.

Pachacuti and Mama Anawarkhi look around together, startled. He moves away to one side.

MAMA ANAWARKHI

What is it, husband?

PACHACUTI

As I feared. The silver miners in Potosí have revolted. This is our latest conquest, a tribe far to the south. They do not follow the orders of our priests and nobles.

MAMA ANAWARKHI

Then you must send the soldiers and make them comply!

PACHACUTI

Now you understand, Anawarkhi. You are a *good* wife.

MAMA ANAWARKHI

It's true, Pachacuti, I am a good wife... But then, I like the silver bracelets just like the gold ones. In fact, the Moon goddess, Mama Quilla, likes the silver ones best of all.

She smiles.
They both bow their heads and exit.

Space Science and Astronomy Script Package #5
"Anthro 121: Folklore and Mythology, Lunar
University, Tycho Crater, A.D. 2101"

Lessons in History and Culture

Educational systems and institutions are anticipated to follow humans wherever they might settle, and that includes in off-world communities and workplaces. This script reflects both the conservative nature of education, in that it teaches about the past, and the transformative nature of education, in that the dramatized, college classroom discussion incorporates new people, their roles and backgrounds, and new understanding of literary, artistic, and historical events. The script depicts the twentieth century as an era in which mythology was used to embody cultural beliefs, and in which mythological creatures were used to convey moral lessons. In many ways, from the vantage point of A.D. 2101, the twentieth century is a bygone historical era whose stories, characters, and cultural motifs are the substance of a college class on mythology—just like the Greek or Roman myths—and would be learned by college students in the same way. One unusual reaction by the college professor in the script is her surprise when she finds she has a student from the "Russian moonbase." One can only imagine the ongoing political relationships between western powers and the Russian and Chinese polities that would give rise to this type of reaction in the year A.D. 2101. Speculating would make a good exercise in political science!

Lessons in Space Science and Astronomy

The substance of the exchange between students and teacher calls for some clarity about exactly how the Milky Way Galaxy is organized by astronomers. Our galaxy is divided according to its most simple structure: the halo, the disk, and the bulge. The Sun is located on the inner edge of the Orion Spiral Arm. The use of quadrants, based on lines seen emanating from the Sun in the plane of the disk, is the traditional way of dividing up the Milky Way. One variant of this scheme is used in the *Star Trek* series—where the center of our galaxy is the center of the coordinates. Other geometric aspects of the Milky Way can be found in the introduction to part 3(B). To measure distances beyond our Solar System, astronomers use the unit of the "parsec," which is an abbreviation for the distance of an object having a *"par*allax," or shift in apparent position over six months, of one arc*second*. It is about 3.26 light years, or 19 trillion miles.

Worksheets for Use before a Script

I. Keywords

Bilbo	Quark (the character)
Chewbacca	Quark (the particle)
Ferengi	Roddenberry (Gene)
Moonbase	*Star Wars*
Mythology	*The Hobbit*
NASA	*The Lord of the Rings*
NASCAR	Tolkien (J.R.R.)
Parable	Tycho Crater

II. Questions for Investigation Before the Script

1. How are the fictional divisions of the Milky Way, in the *Star Trek* series, different from the real divisions of the Milky Way, as used by astronomers today?
2. How big is our Milky Way Galaxy? What units are used to express the size of the Milky Way, and what are its dimensions?
3. There are many images of a "moonbase" on Earth's Moon. What features do you believe are the most realistic, in order to make a moonbase habitable for human beings?

Discussion Questions for Use After the Script

1. The students in the anthropology class had some trouble getting their facts straight. Did the stories in the *Star Wars* series take place in the Milky Way, or not?
2. Schools and universities are present in many human settlements. How realistic do you believe it is to have a university on Earth's Moon? Does it make sense? Why or why not?
3. Tycho Crater is one of many craters on Earth's Moon. What are these craters, and what are they caused by?

Costumes and Props for Inexpensive Productions

In this script, a variety of space costume pieces can be used to suggest life on a moonbase. The script uses different colored baseball caps to distinguish the students, and these are easy to obtain. However, the tricky aspect of a depicted classroom on the Moon is that it takes place in one-sixth of Earth's gravity, and participants must use either a different gait to move and walk, or they have weighted boots, or both. This is difficult to dramatize in a short script, but it may be worth attempting, if for nothing else than the added value of humor for the audience. Humor does help people to remember facts, figures, and lessons.

Introduction to the Script on a Possible University Class at an American Moonbase, A.D. 2101

The tension in this script derives from students' anticipation of an upcoming quiz in a college course on mythology. However, this tension is modulated by humor throughout the script, while students confound one "historical fact" after another. One student confuses "NASA" and "NASCAR," without any other student correcting him. Another student mixes up the character Quark, from the *Star Trek* series, with Bilbo from Tolkien's *The Hobbit*. Yet another student mis-quotes the Alpha, Beta, Gamma, and Delta quadrants from the *Star Trek* series, as "Alpha, Butter,

Camden, and Dexter," suggesting all manner of confusion about the first four letters of the Greek alphabet, not to mention *Star Trek's* Milky Way quadrants! To us in the twenty-first century, it is amusing for people to confuse these characters and institutions, many of whom we know so well. However, the authors found that the biggest laugh during their production of the script came at the point in which one student claims that Jesus made parables "simple to understand" for a personal reason. The character states: "There's a reason Jesus made them simple—for people *like you!*" If this line is expected to cause a negative reaction in the assembled audience, the line can be changed or omitted. However, it is noteworthy that it created the biggest comic reaction during a script that is very subtly humorous throughout. The audience laughed heartily.

Script
"Anthro 121: Folklore and Mythology, Luna University, Tycho Crater, A.D. 2101"

Characters:
STUDENTS in a college anthropology class, "Mythology and Folklore," at Luna University, "A" Dome, American Moonbase, Tycho Crater [can all be played by one person]
PROFESSOR CROCKETT

> Place: Interior, at normal moon gravity (humans and dogs in weighted booties).
> Location: College classroom, Luna U, "A" Dome, American Moonbase, Tycho Crater.
> Time: Morning, just after 10 o'clock, in artificial day.

A [SLIDE OR SIGN] gives the year, time, and place:

"A.D. 2101, just after 10 in the morning, in artificial day. Luna University, 'A' Dome, American Moonbase, Tycho Crater."

PROFESSOR CROCKETT

In front of a class of students.

Ready for review? Remember, each quiz is worth ten percent of your final grade. This is an important period for terran mythology, so get your facts down and be prepared to write on at least one essay question. ...Who wants to start?

STUDENT IN BLUE BASEBALL CAP

Hand raised briefly, then he begins.

I've got a question. What I don't understand is why Quark went on his walkabout, at all–

STUDENT IN RED BASEBALL CAP

–Quark? He was a Ferengi. You mean Bilbo? He was the Hobbit.

STUDENT IN BLUE BASEBALL CAP

Quark, Bilbo—they're all the same! Weird little guys that are supposed to be us, I *guess*... some kind of *symbols*... I mean, I don't understand all these *strange... little...people...*

PROFESSOR CROCKETT

Well first, let's get the facts straight. Quark was a character from the Star Trek series. Bilbo was the Hobbit in books by J.R.R. Tolkien. Can anyone tell me the biggest difference between those two sets of stories? ...Yes?

STUDENT IN YELLOW BASEBALL CAP

Tolkien wrote for children. Roddenberry wrote the first Star Trek series for adults.

PROFESSOR CROCKETT

Right!

STUDENT IN RED BASEBALL CAP

–Wait! I've known adults who read *The Hobbit* and *Lord of the Rings*—

PROFESSOR CROCKETT

Of course! The themes are universal and they transcend generations—themes like warfare, heroism, and the Hobbit's own growth as a self-conscious being. And, the story is known to reflect the author's experiences in World War I, although it's supposed to be in some murky time "Between the Dawn of Færie and the Dominion of Men"...as some have written...

STUDENT IN BLUE BASEBALL CAP

Still, why all these *weird beings* in all this mythology? Like...What about Chewbacca? He's taller than a lot of guys, but he's *a dog,* or something... He thinks like us, but he doesn't even *speak!* I don't get it. It totally *weirds* me out!

PROFESSOR CROCKETT

Could somebody answer that? It's a good question. Why all these half-human, half-animal beings? I'll give you a hint: We found the same thing in Earth's Greek mythology, too. Remember?

STUDENT IN YELLOW BASEBALL CAP

I think it has something to do with *parables*–

STUDENT IN BLUE BASEBALL CAP

—What? What are *they?*

STUDENT IN YELLOW BASEBALL CAP

You know, like the simple little stories that Jesus teaches in the Bible.

Laughing.

There's a reason Jesus made them simple—for people *like you!*

STUDENT IN BLUE BASEBALL CAP

Ouch! ...But anyway... Jesus? The Bible? I'm Buddhist. We don't read the Bible.

PROFESSOR CROCKETT

Wait, wait, wait a minute. Let's take it easy... So tell me, why did Jesus make the stories simple?

STUDENT IN YELLOW BASEBALL CAP

So people would remember *the lesson.* It's easier if the story's simple... but it's also easier, uh, *to digest,* I think.

PROFESSOR CROCKETT

Right! So what you're saying is that mythology has some practical and important purposes, aren't you?

STUDENT IN YELLOW BASEBALL CAP

I guess. Yes.

STUDENT IN BLUE BASEBALL CAP

I don't see *anything* about Star Wars' Chewbaccca that's got anything practical to do with my work refueling rockets coming in and out of 'D' Dome! Zip! Nada!

PROFESSOR CROCKETT

Well, sometimes, it's easier to see a moral lesson if the character is *not* human. People sometimes have a hard time integrating moral lessons when they hit them in the face. You understand? Yes?

STUDENT IN BLUE BASEBALL CAP

I guess....Yeah, I get it: Don't call out somebody's mistakes too loud, or they're likely to pop you one! ...That makes sense.

PROFESSOR CROCKETT

It's one of the most important functions of mythology, remember? To teach moral lessons... But, let's turn to something else. Any other questions right now?

STUDENT IN YELLOW BASEBALL CAP

Hesitating.

Yeah, I was wondering about something...

PROFESSOR CROCKETT

Yes...?

STUDENT IN YELLOW BASEBALL CAP

Did NASCAR come up with the four galaxy quadrants, or was that all mythology, too? You know—Alpha, Butter, Camden, Dexter... Is that right?

PROFESSOR CROCKETT

She hesitates and looks down, trying not to laugh.

Well... First, I think you mean Alpha Quadrant, Beta Quadrant, Gamma Quadrant, and Delta Quadrant. They're from the Star Trek series. They're the quadrants of the Milky Way... and, they're the first four letters of *what* language? Anybody know?

STUDENT IN RED BASEBALL CAP

Gotta be Latin.

STUDENT IN YELLOW BASEBALL CAP

Nah, it's Greek. My Dad's Greek. He uses those words sometimes.

PROFESSOR CROCKETT

Then, of course, I think you mean NASA—the National Aeronautics and Space Administration — not NASCAR... Most of you are American, yes?

STUDENT IN RED BASEBALL CAP

I'm from Chile—the country, not the food.

STUDENT IN YELLOW BASEBALL CAP

I'm from SoCal—Southern California!

STUDENT IN BLUE BASEBALL CAP

I was born in the Russian Moonbase—

PROFESSOR CROCKETT

–You *were?* How'd you get *here?*

STUDENT IN BLUE BASEBALL CAP

Stole a rover. Came over. They let me in.

PROFESSOR CROCKETT

Wow. That's quite a story. Gracious...

STUDENT IN BLUE BASEBALL CAP

It's okay. I like it here.

PROFESSOR CROCKETT

Well...good... But let's get back to the review. Does anyone know if NASA and other astronomers use the Star Trek coordinate system for the Milky Way Galaxy?

STUDENT IN RED BASEBALL CAP

I think they do.

STUDENT IN YELLOW BASEBALL CAP

Yeah, I agree.

STUDENT IN BLUE BASEBALL CAP

Nah, *they do not!* Otherwise, my orders wouldn't read the way they do! When I refuel a rocket, it states on the manifest where the rocket's heading, and that's *in degrees—in two coordinates—*not in quadrants. You know what I mean?

PROFESSOR CROCKETT

So! Your work has something to bring to our discussion after all! Great! Do you want to explain that to your classmates?

STUDENT IN BLUE BASEBALL CAP

Sure. But first, tell me what Chewbacca has to do with morality! I *still* don't get it!

Space Science and Astronomy Script Package #6
"Escape from the Milky Way: A Hyper-Velocity Star Makes a Run for It"

Lessons in History and Culture

Like other scripts in this volume that take place far into the future, the characters interact with technologies that are not yet in existence. The two characters in this script both describe a probe that has been travelling faster than the speed of light by virtue of an Alcubierre warp drive, whose feasibility is now being researched by NASA (*cf* Alcubierre 1994).

The humor in this script is very different from the subtle humor of the college classroom in Script #5. Technically, the lines in Script #6 could be delivered straight, without any humor at all. However, this would be missing a good opportunity.

The humor in this script derives from the juxtaposition of an exaggerated "country accent" or "rural accent," with the (urban) formality of the announcer's voice and bearing. Indeed, the mission flight director in this script likens astronomers' fondness for data to her horse's fondness for oats! She refers to the location of her horse as "back home," suggesting that the place where she works is more formal and urban than her native home or home state. Add to this, the fact that her horse's name is "Buck Rogers," and the details further confound the urban/formal vs. rural/informal difference, which is expected to survive to the late date of this script, A.D. 2520. While Earth is expected to be far more urbanized five centuries from now, there will remain extensive rural areas and their differentiation from urban areas may be even more exaggerated than today.

This script is a prime example of the conservative nature of culture. It usually changes very slowly, especially among rural people, and to a lesser extent, among rural people who migrate to urban jobs. Language accents, vocabulary, and special inflections perpetuate many rural characteristics in urban areas, often without any detriment to the urban work performance of the individuals who keep older language customs.

Spaceport America is a real place near Las Cruces, New Mexico. It broke ground in 2009, and one can only anticipate the vague lines of how it will develop in the future. Students might research its planned development. It should be noted that, while Las Cruces is not a major urban area now, it is a central place for the surrounding rural areas, and it may well become far more urbanized if and when Spaceport America becomes a major location for rocket transportation and commerce, or perhaps a space elevator.

Lessons in Space Science and Astronomy

The center of our Galaxy is completely obscured by interstellar dust, and so it is invisible with probes at the shorter wavelengths, like ultraviolet and visible light. One of the brightest patches in the Milky Way, just above the teapot-like "spout" of the constellation Sagittarius, is called "Baade's Window." Here, the dust is less dense, but it is still present, and besides, it is not quite at the galactic center. It would be a great accomplishment to send a probe to the center of our galaxy, even if it took generations of travel because the distance is 27,000 light years!

This script envisions the future use of the Alcubierre drive to reach the center of the galaxy with a means of traveling faster than the speed of light. Theoretically, the Alcubierre drive does not violate Einstein's special theory of relativity because a spaceship or a probe would travel inside a bubble of spacetime. While speculative, we have noted above that NASA has taken steps to detect the spatial distortions from an expansion and contraction of spacetime, which are needed for the drive to achieve "warp speed." This may not be quite as unimaginable as once thought.

Figure 5. A Runaway Star named Zeta Ophiuchi. (Credit: NASA/
JPL-Caltech)

For those of us who are familiar with the constellations in the
night sky, it can be a surprise to learn that all stars are in relative
motion to each other. Our Sun must have moved a considerable
distance from the open cluster of stars in which it was formed.
Some stars in the Sun's neighborhood are moving at relatively high
speeds vertically through the disk of the Milky Way. They were
formed in the halo of our galaxy and now are just "passing by" our
Sun. We can also observe a few stars in the halo that have been
given enough velocity to escape the gravitational pull of the Milky
Way, itself. When a star achieves this high velocity, it breaks free
and becomes a "runaway star." Figure 5 pictures the runaway star,
Zeta Ophiuchi, which is traveling at such a high velocity (from
right to left, as the picture is viewed) that it creates a curved shock
front.

Worksheets for Use before the Script

I. Keywords

Alcubierre-powered

Binary star system

Black hole; supermassive black hole

Broadcast booth

Cape Canaveral

Decennial

Gravity; gravitational

Hypervelocity star

Mission flight director; mission
   control room

NASA

Off-world

Optical

Probe

Public relations

Runaway star

Spaceport America

Spacetime; bubble
   in spacetime

Supernova

Ultraviolet light

Wallscreen

Warp speed

Xrays

II. Questions for Investigation Before the Script

1. Do you think that either a broadcast announcer or a mission flight director would need to have training in astronomy? If yes, why? If no, what are the fields in which they would need expertise?
2. Why do astronomers sometimes survey a star or planet with different wavelengths of light? What is the name of the field of expertise that analyzes different wavelengths of light?
3. Do scientists understand what gravity is, and how gravitational pull operates?

Discussion Questions for Use After the Script

1. Why did the probe of the Milky Way's center have to be inter-generational? Why couldn't it happen in a single generation?

2. Space travel at warp speed has not yet been achieved, and some scientists believe it can't be reached because of which one of Einstein's theories?
3. The mission flight director speaks of "a bubble in spacetime," in which the probe had been travelling. In general terms, what do you think "a bubble in spacetime" is?
4. Why have the images sent by the probe become "dustier" over time?
5. What is the connection between the supposed supernova and the runaway star?
6. How could Paul and Pam "catch a glimpse" of the black hole?

Costumes and Props for Inexpensive Productions

This script takes place in the future, in A.D. 2520, so it is difficult to anticipate which of the currently expected, futuristic clothes and accessories will survive. The clothes worn by the mission flight director should be comfortable, informal, and perhaps conservative—as she is a government (or corporation) employee. The garments worn by the broadcast announcer might be somewhat more formal, stylish, and/or outlandish, because he is appearing on a widely-broadcast show of humanity's first true sighting of the center of the Milky Way Galaxy. He is both a communications expert and an entertainer, like broadcast announcers are today.

Most important in conveying the futuristic aspect of this script will be sleek headsets or microphones. These are available online from many vendors in all kinds of styles and colors, and at a wide variety of prices. There are also "toy" versions that will serve quite well.

Introduction to the Script on a Possible Inter-Generational Probe to the Milky Way's Center, A.D. 2520

The script will play well to a backdrop slide of a mission control room with consoles, screens, and/or maps of the probe's progress. A "broadcast booth" separates the announcer from the mission flight director. To suggest this, the two characters might appear on different sides of a stage or facing away from each other. Students should be reminded that these roles exist today, and, while education in science and astronomy would be helpful to filling these two occupations, they may not be absolutely necessary. This might also apply in the future when scientific discoveries are expected to be an even more "front page news."

The humor in the script is its most difficult feature because it is "broad humor," as in slapstick, and it is sometimes difficult for people to play. The accents and bearings of the two characters are exaggerated on purpose. This type of humor contrasts markedly with the serious nature of the ongoing work, but that juxtaposition is part of the script's appeal. It is recommended that students rehearse their spoken parts perhaps more than in other scripts because the language accents and timing-dependent "jokes" are important to the staging of its humor.

Script
"Escape from the Milky Way: A Hyper-Velocity Star Makes a Run for It"

Characters:
PAUL CHANDLER, Public Relations Spokesperson for NASA, at Spaceport America
PAM PLOVER, Mission Flight Director, Inter-Generational Probe Project, NASA

Place: Interior.
Locations: Broadcast Booth (Paul) and Mission Control Room (Pam) of Spaceport America, Las Cruces, New Mexico.
Time: Afternoon, 8 February, A.D. 2520.

A [SLIDE OR SIGN] gives the year, time, and place:

"A.D. 2520, February 8. 1:30 p.m. Locations: Broadcast Booth (Paul) and Mission Control Room (Pam), Spaceport America, Las Cruces, New Mexico."

PAUL

He is wearing a small headset and speaking into another large, sleek microphone. Behind him is a wallscreen. He speaks in an announcer's voice and tone.

Good afternoon, ladies and gentlemen. We are broadcasting live from Spaceport America in Las Cruces, New Mexico, for the decennial report on humanity's First Inter-Generational Probe. This is a special one, as I'm sure all of our listeners know. In these upcoming, recorded images, we may well catch our first glimpse of the center of our galaxy—the Milky Way! At ten year intervals, the human populations on Earth and in our off-world territories have been treated to sights from our First Inter-Generational Probe, now nearing the center of our galaxy. Two hundred and twenty years ago this month, we sent a probe to record the black hole that, we believe, lies at the center of the Milky Way Galaxy.

The past several pictures have been dustier and dustier—we expected that, didn't we, Pam?—and today's report will show images as the probe begins to pass directly over the center of our galaxy.

What do we know, Pam? What do we expect that Voyager X will show us? Do we have any idea, at all?

Ladies and gentlemen, meet Pam Plover, Mission Flight Director for NASA here at Spaceport America. It's not the same as Cape Canaveral, is it, Pam?

PAM

She is seated at a console in Mission Control, wearing a sleek headset with its own mic. She speaks with an American, country accent.

Thanks, Paul. Well, it sure is different from Cape Canaveral. We were all sad to see that historic venue sink beneath the waves, but Spaceport America more than makes up for it...

And no, we're not exactly sure what Voyager X will reveal today. Just the fact that it has been traveling at warp speed for over two hundred twenty years—

Cheers drown her out from the other members of the Mission Flight Control Team.

We're mighty proud, as you can tell, of *all* our Alcubierre-powered vehicles, which were first developed right here at NASA. Today, we take special pride in the accomplishments of all the members of our Mission Control Team through the years. Of course, no one is alive today that saw Voyager X launched as the First, warp-powered, Inter-Generational Probe, to glimpse with optical instruments the center of our very own Milky Way. So we're proud, and mighty excited too, aren't we, guys and gals?

Cheers erupt in response from the living members of the Mission Flight Control Team.

PAUL

Pam, give us a summary of what your team has been viewing in the optical recordings leading up to today's special fly-over of the *Center of the Galaxy.*

PAM

Well, Paul, we've seen dust—lots of swirling dust—but through those clouds we've been able to view layer upon layer of stars! Our astronomers have been having a Field Day, I'll tell you! As you know, they've been trying to map the densely-packed stars at the galactic core with xrays and ultraviolet light, but it's just too darn dense! And too dusty! Now, they have good material they can use for decades to come!

PAUL

That sure is exciting, Pam. Always good to keep the astronomers, happy—

PAM

Yes, indeed. A busy astronomer is a happy astronomer. They feed on large quantities of data, just like my horse back home—Buck Rogers. He lives off *lots and lots of oats!*

PAUL

It's a good comparison, Pam. But, tell us now what to expect today. There's some special news I was reading from NASA yesterday, about the possibility of seeing a 'runaway star'! Is that something like a 'runaway bride,' Pam? Tell us what a runaway star is, in plain simple terms, of course...

PAM

Well, Paul, you see there are these Hy-per Vel-o-city Stars out there. They go so darn fast, they break the gravi-ta-tional pull of the entire Milky Way. They form when a bi-nary star sys-tem—that's two, you know—is pulled toward the su-per-mas-sive black hole at the center of the galaxy. We think—*we think*—that one of the stars could explode in a su-per-no-va explosion.

So what happens to the other star? We think the other star achieves an enormous speed with the su-per-no-va ex-plo-sion and is *ejected* from the galaxy. *Thrown out!* It breaks the pull of the galaxy's own gravitational pull and becomes... *a runaway star!*

PAUL

Wow, Pam, you sure make it exciting.... So...what are we seeing on the view-screen now?

PAM

We've got just about a minute before Voyager X should be directly over the black hole. We've calculated very, very carefully, and we believe, if we're real, real lucky, that we'll also catch a glimpse of that runaway bride—I mean, runaway star! Let's look...

PAUL

I see the dust clearing a little...

PAM

Look at those densely packed stars, Paul...

PAUL

Pam, what's that dark object we can just... catch a glimpse of... every once in a—

PAM

—Oh, wow, that's not an object! That's the hole! That's the black hole! See it? It's black as tar! You can just catch a glimpse of it! Isn't that *beautiful!*

PAUL

If you say so, Pam, but that thing looks scary to me...

PAM

Now wait, let's see if we can see an explosion. That would be the su-per-no-va explosion—where a star explodes, you know.

PAUL

I'm waiting, Pam, along with the rest of humanity...

PAM

Oh! There is it! The astronomers were right! Hot dog!

PAUL

But where's the runaway—

PAM

There it is! Wow! It looks like it's coming straight toward the probe! My good-ness gra-cious! ...It's gonna pass us by! Now, I think... Oh, no!

PAUL

What, Pam? Tell us what's happening.

PAM

We're falling!

PAUL

You mean, the probe is falling?

PAM

Yes! It's falling toward the black hole! The gravi-ta-tional pull of the black hole must have disrupted the bubble in spacetime that our probe has been travelling in! Oh, no!

PAUL

...*What...?* Pam... Where are the rest of the images? *Pam, tell our viewers what's happened!*

PAM

I think that Voyager X was sucked into the black hole and it's gone, Paul. It's been eaten by the black hole. Voyager X is dead. Oh, my goodness, this is the end of the Inter-Generational Probe Project!

PAUL

The end?

PAM

The end.

C. Stars Scripts

Science usually starts with the classification of objects, and the science of stars is no exception. The main way astronomers classify stars is through the spectra of light they emit. When a pioneer of stellar classification, Father Angelo Secchi, S.J., looked at his drawings of stellar spectra, he found they were not all the same, but fell into distinct classes to which he gave the Roman numerals I, II, III, and IV. These were refined and expanded into the letter sequence of OBAFGKM. The ordering of the letters looks peculiar, but the sequence follows the progressive change of features in their spectra. These features were later correlated with the decreasing surface temperature of the stars producing the spectra.

For main-sequence stars like our Sun, which are fusing hydrogen into helium in their cores, the sequence is also one of decreasing mass. The hottest, most massive stars are the quickest to go through their evolutionary cycle. Only when the sequence gets down in temperature and mass to be the "late F-type" or "G-type" star is the star's stable, main-sequence stage sufficiently long for life to develop on accompanying planets. For K and M stars, their stable lifetimes are even longer, so these too are proposed as possible hosts for planetary life. Not surprisingly, our Sun is classified as a G2-type, main-sequence star. It is certainly friendly to life, providing a planet that orbits at a distance where water can remain in liquid form.

Before nuclear synthesis was found to be the power source of stars, the letter sequence of stars (OBAFGKM) was understood as an evolutionary sequence. The stars were thought to cool down as their supply of whatever fuel—maybe coal—gave out. However, nuclear synthesis is much more resourceful than the combustion of coal. The product of hydrogen synthesis, helium, becomes the fuel for the next stage of the evolution of stars, and so on. The later stages of stellar evolution result in an enormous increase in the size of stars, and therefore, an increase in their brightness.

Astronomers also classify galaxies, but they do this according to how they look visually, although their spectra broadly follow the "tuning fork diagram" sequence developed by Edwin Hubble. To the three broad classes of galaxies—elliptical (Figure 6), lenticulars, and spirals (Figure 4)—the class of irregular galaxies are added. The classification sequence of galaxies was not initially an evolutionary sequence. However, it can be used to describe how the merger of spiral galaxies from the early universe formed present-day elliptical galaxies. The lenticulars are probably elliptical galaxies stripped of their gas and dust, perhaps by further mergers.

Figure 6. The Sombrero Galaxy, an elliptical galaxy showing a dust lane around its outer edge. (Credit: AURA/STScI /NASA)

At distance scales far beyond those of the clustering of galaxies, we find that all galaxies are receding from each other. No single galaxy is privileged, so there is no center of the universe. Everything is expanding from the unimaginably small dimension of "spacetime" at the Big Bang. The fate of all these galaxies depends on whether the universe is sufficiently dense to allow the mutual gravitational attraction of galaxies eventually to overcome their expansion. From present observations of Type Ia supernovae, it seems that not only is this incorrect, but that the very expansion rate of the universe is increasing. All galaxies are receding from each other at an increasing rate. The universe is a mysterious place.

Space Science and Astronomy Script Package #7
"Early Humans at Kenya Cave: Middle Stone
Age Star Charts, 150,000 Years Ago"

Lessons in History and Culture

"Early Humans at Kenya Cave" is one of two scripts in this volume that are purely dramatic. Most others are humorous or light-hearted. This story is serious because it concerns changing climate, the search for food, and the pressure of natural selection on an early human population. During the Late Pleistocene Epoch, Africa was subject to changing climate and sometimes early human populations had to move. The script provides an intimate, fictional account of one such instance, based on actual archaeological finds of etched stone, shells, and ochre. There is drama and tension in this script, while the two characters decide what to do.

Archaeological remains from the Middle Stone Age in Africa suggest that members of the genus Homo had evolved at least some of the rudiments of science, religion, and art. For example, sites reveal the following: bedding made of insecticidal plants, which could show knowledge of a pharmacopeia; same-sized shell beads that are carefully and uniformly perforated, which show wear patterns from being suspended around the neck; purposefully defleshed skulls, which could show a reverence for the dead; carefully worked bone tools and bladelets that were hafted with a type of mastic; signs of long-distance trade; red (and other colors of) ochre for the processing of paint; and purposefully etched stone and ochre pieces with patterns made by sharp tools. All of these artifacts hypothetically constitute the "external storage of symbols," or the external storage of cultural patterns outside the early human brain.

We place Kenya Cave on the eastern coast of Africa, south of the area in which so many early human remains have been found, in present day Ethiopia. For our fictional archaeological site, we propose a cave habitation that dates to around 150,000 years ago. While the artifact used in the script is an etched "stone," other media have been used, such as shells and ochres (also a type of pigment, when ground). Artifacts like these are often found with other implements, such as grinders, scrapers, and burnt bone, and for some later sites, beads, and bone tools. Sites span Africa and Eurasia, all the way to China. Some are identified for certain species, such as *Homo erectus*, *Homo neanderthalensis*, and *Homo sapiens*.

Our fictional Kenya Cave in East Africa shows evidence of hominins who were adept at making tools, but it is not clear which species made them. The hominins at Kenya Cave may have been anatomically modern man (AMH), at 150,000 years ago, but it would not be certain until fossils were found with the tools. They would definitely be members of our genus, *Homo*, since no other genus is known to make tools.

Lessons in Space Science and Astronomy

Charts with the positions and brightness of stars have been depicted and used in navigation, both by land and sea, since humanity was able to carve or draw. An early, portable chart with a sighting arm is called an "astrolabe". Its sighting mechanism evolved into the sextant. Professional astronomers today often find their charts online in the Aladin Sky Atlas, which combines many sources of information, or object catalogs, together. These have now become astronomers' own "external storage of symbols."

The man in Script #6 refers to "twinkling stars." Stars seem to twinkle because the varying layers of the Earth's atmosphere diffract the light, very slightly, first one way, and then the other. The effect is more pronounced for a point-like source, a star, than for an extended source, a planet.

The sky viewed by early humans at Kenya Cave will have changed a little in the past 150,000 years. Over long periods of time, stars move in relation to each other (called "proper motion"), and so constellations change their shapes over time.

Worksheets for Use before a Script

I. Keywords

Band of hominins	*Homo sapiens*; and *Homo sapiens sapiens*
Engraved ochre	Spirits
"External storage of symbols"	Star chart

II. Questions for Investigation Before the Script

1. The encounter in this script is supposed to take place 150,000 years ago. Which species could have made the implement in the script?
2. What does the term *"Homo sapiens"* mean? Which typological (biological classification) division does "Homo" refer to, and which does "sapiens" refer to?

Discussion Questions for Use After the Script

1. The man in this encounter uses the phrase "star chart of the land." What word would we normally use in place of this phrase?
2. The woman says that the "herds are more plentiful to the north." What does that imply about their position on the globe?
3. The man uses the term "twinkling stars." Why do stars "twinkle"?
4. The term "external storage of symbols" is often used by anthropologists now. What was the example of "external storage of symbols" in this script?
5. What emotions predominated in this script? Why?

Costumes and Props for Inexpensive Productions

It is very probable that the individuals at Kenya Cave wore clothing of some kind, if for no other reason than to protect themselves from the elements. Hominins who had the sophisticated tools used to process ochre, also probably had the awl, which is a tool to puncture holes and sew skins together. Since weaving and fabric would come along much later in the cultural evolution of humans, the authors supposed that the characters at Kenya Cave wore skins.

With this in mind, we obtained very inexpensive fake fur fabric and fashioned cheetah-like fur pieces to dress for the Kenya Cave script. The skins were worn over same-color pants and turtlenecks to approximate human skin. Since shells were also present, we supposed that one of them wore a shell necklace. That was obtained from an old collection of jewelry, but similar shell necklaces can be obtained quite easily online or at the beach. One hundred fifty thousand years later, humans are still fond of wearing shell necklaces!

We used two props for the Kenya Cave script. We found a night light that was battery operated, which resembled a campfire. We turned it on for the conversation between Seer and Em around the campfire, which added a truly "camp" feeling to the script—in all meanings of that word. We fashioned a piece of etched stone from a chunk of plaster-of-paris that we painted reddish brown. A similarly-shaped piece of rock or granite could work just as well.

Introduction to a Script on the Discovery of Star Patterns by Early Humans, 150,000 Years Ago

In this script, a piece of etched stone is used as a kind of map (and another piece is spoken about as a kind of chart of the stars). No one knows the exact purpose of comparable, real finds in places across Africa and Eurasia. We have simply guessed. There are as many possible uses as we, their near relatives, can imagine. It is difficult to believe that human relatives who were as good hunters as the individuals who left these archaeological remains, did not study the sky, notice the movements of stars, and begin to try to predict the seasons from the sky. Their lives depended on it. At many times throughout the Middle Stone Age, human ancestors held precariously to life.

Script
"Early Humans at Kenya Cave: Middle Stone Age Star Charts"

Characters:
SEER, medicine man for a band of about twenty-five early *Homo sapiens* living at Kenya Cave, on the coast of eastern Africa.
EM, a woman who is the daughter of the agèd headman.

Place: Exterior.
Location: A man and a woman are seated at a campfire, warming themselves.
Time: In the evening, after dark.

A [SLIDE OR SIGN] gives the year, place, and time:

"Approximately 150,000 years before the present. On the eastern coast of the African continent. 9:30 p.m. in the evening."

SEER

The air is cold tonight.

EM

Hm. And last night, too.

SEER

It is the wrong time of the year to have
cold nights.

EM

Hm...maybe, but, the stars are very bright.
They are bright when it's cold.

SEER

The spirits are restless tonight, Em.
Can you hear them calling in the wind?
Can you see them speaking to us in the
twinkling stars?

EM

No, only you hear and see such things,
Seer. I hear the wind. I see the stars. I
wouldn't know they were spirits...
without you.

SEER

He looks at her and smiles fondly.

You are a good woman, Em. Smarter than
most! And braver than most! You have
gone to meet the Others to the north
of us...

EM

Hm. I went with my father and the other men, to help them with the skins...We were hungry, so we followed the animals. That's not brave. That's hungry.

SEER

But they say that you went out first to meet the Others, that you walked with your hands open to the headman of their group, and that you signaled with your hands, and they understood...You are brave, Em.

EM

The herds are more plentiful to the north, Seer. Have your spirits told you that?

SEER

Your father believes we should wait here, that we should be patient, Em. There has always been food here from the sea... although it is harder to catch... The weather will improve. We have had bad years before.

EM

Hm. He is an old man, Seer. We must leave if we are to live. You must go. You must find us a better place and make friends with the Others. You can do this, Seer. I've seen you talk with people. Talking with the spirits is useful, but talking with people is more useful. You must go.

SEER

He hesitates, and stirs the campfire.

Not without you.

EM

She shakes her head.

I am too old for you, Seer. I was five winters and five summers when you were born.

SEER

You are not too old for me, Em. Remember, I have seen how you talk with people, too.

He laughs.

EM

Why are you laughing?

SEER

You might be as good as me!

EM

Hm. Foolish man...I will not go with you, but I will show where to go and how to get there. And...if you are very lucky...I will wait here for you and, hm, consider your...offer.

SEER

How can you show me where to go if you don't go with me? I don't understand, Em.

EM

She draws an etched piece of stone from her cloak of skins.

Take this, Seer, and use it to find the Others.

SEER

He takes the piece and peers at it.

This is like the pictures I make of the positions of the stars.

EM

She turns quickly to look at him.

On a piece like this?

SEER

Yes, I've seen others making the marks. I did the same.

EM

You are clever, too, Seer...But, this is not a picture of the positions of the stars. This is a picture of *your* position—the signs you follow to find the Others.

SEER

Oh! It's a star chart *of the land!*

EM

Hm. Yes, in a way. It tells you which way to go when you reach each of the tall peaks. I remember them, and the direction we came back home.

SEER

He shakes his head.

I cannot...

EM

You *must*. The nights are too cold here! Too many of our babies are dying, Seer. You must take this and find another place with Others who will welcome us...*Talk* with them, Seer.

SEER

He stares at her for some moments.

I will go...but you must promise me that you will be here waiting.

EM

I will wait for you. Just find us a way to survive! You can do it, Seer! Just follow the star chart of the land, and talk to the Others the way you do so well.

Space Science and Astronomy Script Package #8
"Orion the Hunter: Star Factory Game Edition 4.0"

Lessons in History and Culture

Astronomy forms the basis for the action in the video arcade game in this script. "Orion the Hunter" refers to a well-known constellation including the star Betelgeuse (Alpha Orionis) to the north, and Rigel (Beta Orionis) to the south. The figure of Orion is often identified from his belt and sword, which includes the Orion Nebula. The belt points to the brightest star in the sky, Sirius (which is actually a binary star system). All of these features make the Orion constellation relatively easy to pick out for first-time stargazers. An entire classroom lesson could focus on all of the interesting facts and processes at work in the constellation we call, "Orion."

The video arcade game based on Orion, and in particular, on the Orion Nebula, where star formation is active, is part science and part fantasy, like most video arcade games. Therefore, the script becomes a good example of distinguishing "good science" from the fantasy used to make the arcade game exciting.

This script, like Script #6, is based on broad humor and regional accent, but it can be played with any chosen accent, or none. It takes place in the near future, in a video game arcade, in a mall, in A.D. 2032. The accent chosen for this script mimics television shows such as "The Sopranos" or "Jersey Shore," which students can research online. It is possible to find voice instructors on YouTube, giving directions on how to produce a variety of English-language accents. The use of accents is often very important in both comedy and drama, because accents often signal social information—one's original language community and one's present language community, one's social status and education, and one's ethnicity. "Anthropological linguistics" is an important field of study that generalizes about variation in language inflection, culture, and society.

Lessons in Space Science and Astronomy

Various aspects of star formation have been mentioned for the scripts in part 3(A) and 3(B). The focus of this script is on the star formation that takes place in dense regions of molecular clouds where component gases and dust can contract under mutual gravitational attraction, and nuclear fusion is turned on in their cores, or, the stars "ignite." The "star nursery" or "star factory" chosen for this script is the Orion Nebula, which contains the mass for about two thousand stars in a region that is twenty light years across.

All the stars in a particular sub-region of the Orion Nebula are formed at the same time, out of the same collapsing molecular cloud. The newly forming stars go through several stages of formation. At first, they appear with a cocoon of outer gas and dust, which is eventually blown away by the radiation of the new star.

The most massive stars (O-type and early B-type) dominate star formation within a particular cluster since they have the strongest stellar winds to "evaporate" nearby molecular clouds. They also go through their life-cycles fastest and trigger new star formation from their resulting supernovae. The most massive of these OB stars will form black holes rather have than neutron stars at the center of their remnants.

Worksheets for Use before a Script

I. Keywords

B-type star	NROTC
Black hole	O-type star
Clouds of gas	Orion the Hunter
Console (n.)	Star nursery or star factory
Constellation	Supernova
Game arcade	Video game

II. Questions for Investigation Before the Script

1. Who or what is "Orion the Hunter"? Why is he called a hunter?
2. Why do young stars have "clouds of gas" surrounding them?
3. How are stars classified? According to what criteria?

Discussion Questions for Use After the Script

1. Tanya refers to Orion's "belt of stars." What is she referring to?
2. Arcade games often combine real science, and fantasy or fiction. Ricky describes two dangers to young stars in the script: Big stars and black holes. Are both of these factual dangers?
3. Do young stars in "star nurseries" or "star factories" actually "blow their clouds of gas away"? What does that mean? Is it unusual or normal?

Costumes and Props for Inexpensive Productions

The authors used very simple costumes for the script in a video game arcade: Baseball caps. The cap for the male character was black, and included an image of a starry night, which was worn "backwards" with the bill toward the back. The cap for the female character was peach color and it had rhinestone sparkles all over it. Other than those caps, the clothing for the script reflects styles of high school students, probably informal shirts and pants, although Tanya might wear a skirt. The action of the script does not rely on clothing, so the simpler, the better, unless one wants to anticipate high school clothing and accessory styles in A.D. 2032! This is always possible.

Introduction to a Script on a Video Arcade Game That Uses Black Holes, and O and B Type Stars

The humor in this script about an astronomy-based arcade game is not simply dependent upon the accents of the characters, but on their gender interaction. The story involves a discussion between two high school seniors who are "boyfriend and girlfriend." While both characters are planning to attend college, one character clearly has a better chance of landing in an occupation that might involve flying fighters and shooting down enemy aircraft—which is exactly the action in the video arcade game. Therefore, the playing of video games is seen as a possible testing ground for manual and visual dexterity, which indeed varies among individuals. Not many make it to the "Top Gun" level of performance of the professional fighter pilot, but some do. One of the characters in the script has a chance to make it to that level, but it may not be the character one supposes at first, and that is one of the sources of tension in the script.

Realization of the differences between the two characters is part of the appeal of the script, especially in the final reveal. If it is possible to show the changing and eventually very high score of the final player of the video game, then that would make a good backdrop for the script. The authors just opted for a final shot of the final high score. Students who know about video production might want to take on making a simple movie or an animated GIF, to show the changing score.

Script
"Orion the Hunter: Star Factory Game Edition 4.0"

Characters:
RICKY, a recent high school graduate, headed for a junior college in the fall.
TANYA, his girlfriend, a recent high school graduate, also headed for college in the fall.

Place: Interior.
Location: Fifth-floor Game Arcade, My America Mall, Trenton, New Jersey.
Time: 9 pm.

A [SLIDE OR SIGN] gives the year, place, and time:

"A.D. 2032, June 25. Fifth-floor Game Arcade, where two recent high school graduates talk at an arcade game called, 'Orion the Hunter: Star Factory.' My America Mall, Trenton, New Jersey. 9 p.m."

TANYA

She is chewing gum vigorously.

Eh, Ricky how's it going? Whaddya doin' up here?

RICKY

At a console, engrossed in a video game.

Hey, Tanya... I'm hangin'...

He smacks the controls hard.

Whammo!

TANYA

Whaddya doin', Ricky? Playin' that silly game?

RICKY

It relaxes me, okay? Besides, it's very educational.

TANYA

Whaddya learnin', Ricky? How to shoot down enemy spacecraft? You think you're going to end up shooting down enemy spacecraft, Ricky?

RICKY

Tanya, gimme a break! Look—I'm learning how to protect myself. They got this star factory, ya know? It's a place where these really big stars *millions* o' times brighter than our Sun are formin' in Orion—

TANYA

—Orion, hunh? Ya know where Orion *is*, Ricky?

RICKY

Sure. It's one of those—you know—constellations up there. Orion's a hunter, that's what he is.

TANYA

He's got a belt of stars. That's how I always can find him.

RICKY

Okay...Well, so Tanya, ya see, when the stars are young, they're protected in their clouds of gas—hidden like, ya know? The object is to zap all their enemies until they get big enough to fend for themselves.

TANYA

You're kiddin' me, Ricky.

RICKY

Nah, it's simple, but it takes some planning. They're more vulnerable, ya see, when they're growing up. When they blow their clouds of gas away, they're not so protected, and they're, like, more vulnerable. There's all kinda things that can get 'em—like other big stars and even black holes—

TANYA

—Ricky, black holes don't move around and attack anything! That's crazy! They just sit there and suck things in with their *gigantic gravity!* There's one at the middle of our galaxy, ya know.

RICKY

You sound like some kind of expert, Tanya. Where'd you learn all that?

TANYA

Shrugging.

I been readin'.

RICKY

I see. Well, like I say, other things can happen to these stars, too, like they can get so big, they blow up! Ka-boom!

TANYA

You mean the star goes supernova?

RICKY

Hunh? You're past me, girl. These stars just blow up if they eat too much. Ka-boom! I love it! But then the game's over and I gotta start again.

TANYA

What's your highest score, Ricky? I'm just curious.

RICKY

I think I did 25,000 one time.

TANYA

Oh, okay—

RICKY

—Tanya, I forgot! Listen, I gotta go pick up my sister when she gets off work. Can I see ya tomorrow?

TANYA

Yeah, sure, I'll be around. Tomorrow's Saturday. Come over about noon. I've got some stuff before that.

RICKY

Preparing to leave.

Whaddya doin' tomorrow morning?

TANYA

I gotta fill out this thing for school. It's for N-R-O-T-C. Ya know, it'll help me pay for school.

RICKY

N-R-O-T-C? What's *that?* Sounds like alphabet soup.

TANYA

Uh... Lemme see... Navy... Reserve Officer Training Corps... NROTC. You know, my Dad's in the Navy. He's a pilot. I'm just carryin' on *the family tradition.* It helps pay, ya know.

RICKY

Wow. Okay. Whatever... Listen, I gotta run. I'll come by for you at noon.

TANYA

Okay, Ricky, take it easy. Hi to your sister.

RICKY

Right, see ya tomorrow.

He leaves.

TANYA

*She looks around carefully and assumes
the controls at the game console. Shaking
her head, she speaks to herself.*

25,000 points... Well, let's just see how
well I do tonight...

*She smacks the controls of the game
rapidly and the counter begins
[SCREEN]. The video game responds,
while she vocalizes her progress.*

Bam!
Bam!
Bam!
Bam!

*The score changes rapidly.
At 1,400,000 points, she stops.*

1,400,000 points! Star Factory Level 21!

Yes! Look at those big babies! Those big
O and B type stars! No black hole's gonna
get my stars! I'll shoot 'em down deader
than a doornail! Wa-hoo!

She smacks the console a final time.
...I gotta go home.

She turns and leaves.

Space Science and Astronomy Script Package #9
"Adventure to Kepler 11757451, Destined
for Exoplanet 'Havre de Grâce'"

<u>Lessons in History and Culture</u>

The action in an adventure to a Kepler exoplanet, in A.D. 2402, takes place between two individuals with very ethnically distinct names—one, Icelandic (originally Nordic), and the other, Irish. Indeed, Thorvald Eiriksson (with an alternative spelling) was the son of Erik the Red and brother of Leif Erikson. He was said to be on an expedition to "Vinland," and the first European to die in North America. Kelly Ann Joyce has the same last name as a famous writer of Irish descent, James Joyce. It is difficult to determine the degree to which the characters carry this cultural history with them, if at all. Ethnicity appears to have little bearing on the occupations they have followed as a science officer (Thorvald) and an agronomist (Kelly Ann) on a starship. Given the vagaries of international travel today, and someday, interplanetary, and even interstellar travel, it is almost impossible to predict how individuals and the ethnicities they represent will sort themselves out demographically by the year A.D. 2402.

Nevertheless, by that year, there should be less sex role-typing, although in the case of this script, the cool, calm, male science officer named Torvald stands in contrast to the emotional and at times, frantic female agronomist and greenhouse manager, Kelly Ann. It is completely up to students how they want to play these two parts, and they may want to change the sex of the students playing the two roles to determine if the script plays differently. The authors stayed true-to-stereotype, with a male playing the science officer's calm role and a female playing the frazzled agronomist. At minimum, the switching of parts should generate a good discussion on sex role and ethnic stereotypes. In this discussion of sex roles and ethnicities, the students might be reminded that the ship's name, "White Amur" is a type of carp that ranges from Viet Nam to

China, where the Amur River is located. The name of the exoplanet, Havre de Grâce, means "haven (or harbor) of grace" in French. If the crew were to wake up the Captain of the starship from hibernation—which they decide to do, at the end of the script—there is no telling how many more ethnicities might be added to the mix!

As with other scripts that are set in the distant future, the adventure to a Kepler exoplanet circling a distant star involves technologies that do not yet exist. There are no starships, no sustaining, onboard greenhouses (although their design is in development at the University of Arizona's Controlled Environment Agriculture Center, and in the literature), no warp drive (although one technology is being researched at NASA today), and no medical technology that promises to keep humans in hibernation safely for over twenty-three years. Nevertheless, all these technologies can be imagined, as can artificial gravity that involves more sophisticated science than spinning a starship to create centrifugal force.

Lessons in Space Science and Astronomy

The theoretical Alcubierre drive, used to achieve warp speed for the starship in this script, is described for Script #6. That information applies to this script, as well.

Selecting a habitable exoplanet for humans is a matter of choosing one that is as close to Earth-like, as possible. The exoplanet should be in a habitable zone (HZ) around its host star. This means that water on the surface can exist in liquid form, and that there is sufficient atmospheric pressure to maintain the water. It should also have a rocky surface rather than a gaseous one like Jupiter. It should have a host star of low enough mass to have a stable, hydrogen-fusing phase that is long enough for life to have formed (found in the late-F, G- and K-type stars), yet not a mass so low that the planet will have one face locked toward the star (found in M-type stars). See the Introduction to part 3(C) for further discussion of stellar types and evolution, and the Introduction to part 3(D) for more on the HZ planets.

Looking at the spectra of exoplanets is currently a challenging science. Sometimes even a high-level haze in the exoplanet's atmosphere seems to conceal the details. On some exoplanets, water, methane, carbon dioxide, carbon monoxide, and elements like sodium have been detected. It is believed that, if oxygen is found in an exoplanet's atmosphere, it would be a sign of chlorophyll-based life. Other biomarkers might be ozone, methane, nitrous oxide, and chlorine. Four centuries from now, detection methods will be much more refined, especially if the observations are made relatively close to the exoplanet in its own solar system. Various types of vegetation should be distinguishable.

Worksheets for Use before a Script

I. Keywords

Agronomist	Hydrolyzer
AI, or artificial intelligence	Instrument panel
Alcubierre drive	Interstellar
Amino acids	Protocol
Artificial day	Spectroscopy; spectroscopic
Artificial gravity	Star system
Bridge (of a ship or starship)	Superterran
Fungi	Universal Translator
Fusion generator	Warp speed; sub-warp speed
Greenhouse	

II. Questions for Investigation Before the Script

1. What are the different views on the feasibility of warp drive? Is it realistic or not?
2. What might the duties of an agronomist on a starship be?
3. Is it possible to create a situation of "artificial gravity" with today's science?

4. How is a "terran" exoplanet different from a "superterran" exoplanet?
5. What could a scientist learn about a planet and its vegetation from spectroscopy?

Discussion Questions for Use After the Script

1. Why was Kelly Ann surprised that the AI awoke just the two of them? Do you think Torvald was surprised, too?
2. How do you feel about using insects and fungi as a source of food on a long interstellar voyage? Why might they both be a good thing to nurture in a greenhouse?
3. Why was the signal offering assistance from "the wrong direction"?
4. Why would there be a "protocol" (explicit rules of behavior) about "Contact"?

Costumes and Props for Inexpensive Productions

There is a variety of "space costumes" for sale online and for a variety of prices. The authors acquired two inexpensive Star Trek costumes—a tunic for an officer and another tee shirt—and then bought two NASA emblems from another vendor. The NASA emblems were sewn on the shirts of the two characters, and both wore black pants. Since the action takes place four centuries in the future, on an interstellar starship with artificial gravity, the characters should be able to move about normally while they try to determine the reason that their AI (artificial intelligence) awoke just them.

Introduction to a Script on Using Warp Technology to Travel to Nearby Exoplanets

The notion of travelling to a relatively nearby (in astronomical terms) exoplanet from which one has successfully received "biosignature data" (possible signs of life, such as carbon dioxide, water, and chlorophyll) is not as far-fetched as it was just a few years ago. There is great interest in exoplanets today, their different types, the kinds of stars around which they orbit, and scientists' ability to determine if they have plant life (see Script #11 in this volume, which shows work on biosignatures four centuries *before* the action in Script #9). The tension in this script derives from the mystery of the AI's waking just the science officer and the agronomist—and not the Captain. The reason becomes obvious by the end of the script, but the two characters decide they should awake the Captain for a reason far more important than determining biosignature data from Havre de Grâce!

Script
"Adventure to Kepler 11757451, Destined
for Exoplanet 'Havre de Grâce'"

Characters:
LT. TORVALD ERICKSSON, Science Officer, USNA Starship White Amur
KELLY ANN JOYCE, Agronomist and Greenhouse Manager

    Place: Interior, Bridge of the USNA Starship White Amur, in artificial gravity.
    Location: 0.5 AU from planet Havre de Grâce, whose star is 2,564.4 light years from Sol.
    Time: Night-phase, in artificial day.

    A [SLIDE OR SIGN] gives the year, place,
    and time:

    "A.D. 2402, *en route* to a warm superterran
    planet at sub-warp speed after almost 23
    Earth years in hibernation at warp speed,
    on an Alcubierre drive-powered starship,
    White Amur. Night-phase, about 4 a.m."

AGRONOMIST KELLY ANN JOYCE

Stumbling in, dazed and confused.

Torvald?

LT. TORVALD ERICKSSON

Engrossed, rapidly using his instrument panel.

Kelly Ann.

AGRONOMIST KELLY ANN JOYCE

No one woke the others? It's just us two?

LT. TORVALD ERICKSSON

Still distracted.

It seems so. I'm trying to figure out why the AI awoke just the two of us. Our systems seem to be checking out, all except for the oxygen level, which has been fluctuating. I don't understand it.

AGRONOMIST KELLY ANN JOYCE

Beginning to focus.

Let me look at the greenhouse panel...

LT. TORVALD ERICKSSON

Hmm. That might—

AGRONOMIST KELLY ANN JOYCE

She scans her instrument panel for the onboard greenhouse.

--Torvald! That's it! ...This is bad! The plants and fungi colonies are all nearly dead! The insect colonies are gone! The bugs are dead!

LT. TORVALD ERICKSSON

...And we thought they'd out-live us all... But that explains it. The greenhouses supply a small part of our oxygen, but when there's little or no output, the hydrolyzers don't function as smoothly.

AGRONOMIST KELLY ANN JOYCE

We'll suffocate, Torvald!

LT. TORVALD ERICKSSON

No, we won't, Kelly Ann. The hydrolyzers can produce enough oxygen for us. That's not the problem. Our problem is *food.*

AGRONOMIST KELLY ANN JOYCE

Plan B.

LT. TORVALD ERICKSSON

Without the insects, it's hard to live on fungi alone. We won't get enough of the right amino acids. You know that!

AGRONOMIST KELLY ANN JOYCE

I know that... Plan C.

LT. TORVALD ERICKSSON

Right.

AGRONOMIST KELLY ANN JOYCE

How long to Harve de Grâce?

LT. TORVALD ERICKSSON

Depends on propulsion. We can't use the Alcubierre drive inside the star system. We won't be able to steer the thing! White Amur might run smack into Kepler 11757451! ...And that's no joke. We'd be toast.

AGRONOMIST KELLY ANN JOYCE

That means we're powered by the fusion generators. So...How long to Harvre de Grâce at sub-warp?

LT. TORVALD ERICKSSON

A couple months, maybe a little more, but we can begin to survey the planet's vegetation right now, with our spectroscopic instruments. The programs should work well, since we're so close. If there's vegetation suitable for eating, they'll find it!

AGRONOMIST KELLY ANN JOYCE

...Did you send a distress signal to Earth?

LT. TORVALD ERICKSSON

Yes, first thing... according to protocol. ...Not that Earth can help us much at this distance...

He turns his back to his instrument panel and folds his arms, thinking.

AGRONOMIST KELLY ANN JOYCE

...Torvald?

LT. TORVALD ERICKSSON

Yes.

AGRONOMIST KELLY ANN JOYCE

There's a signal coming in. That was fast!
Maybe Earth has a faster way to detect our
signals now, through the other interstellar
colonies. We've sent out a dozen ships of
settlers. Maybe one of them picked up our
distress signal first!

LT. TORVALD ERICKSSON

He turns around and looks at his screen.

This is crazy... It's coming from the
wrong direction! We're supposed to be
the farthest colony yet. How can the
signal come from farther out?

*[SCREEN] shows incoming writing,
flickering and dithering, gradually
clearing up, as the Universal Translator
converts an unknown script to English.*

It reads...

Your distress signal has been received. We
would like to be of assistance. What is
your exact speed and position?

AGRONOMIST KELLY ANN JOYCE

Torvald? *Where* is that coming from?

LT. TORVALD ERICKSSON

*He shakes his head, and turns and looks
at her.*

It's coming from Harvre de Grâce.

AGRONOMIST KELLY ANN JOYCE

We have Contact... We have *Contact?*

LT. TORVALD ERICKSSON

Calmly.

I think... It seems so. We have Contact. It
had to happen sooner or later, Kelly Ann.
I guess we should wake the Captain. He's
our official Ambassador... according to
protocol.

D. Planets and Exoplanets Scripts

Planetary science includes everything you want to know about how you got here and whether you might have intelligent counterparts beyond our Earth. It is very comprehensive and very interdisciplinary.

It was obvious to early humans on Earth, who were often favored by clear dark skies that, while most of the stars kept fixed places in relation to each other, there were a few that "wandered." Later, we learned that these are not true "stars," but parts of our solar system—the planets and our Moon. They are very much closer to Earth than the seemingly fixed stars.

We know something of how our Sun was formed in the company of ten to twenty thousand other stars, from the gas and dust in an element-rich giant molecular cloud (see Script Package #8). Not all of the collapsing, slowly rotating gas and dust of our early solar nebula went into the central star. Some remained in a debris disk around the new Sun, and as it cooled, the debris of dust and gas accreted, or maybe condensed from local swirls, into the other components of our solar system: the planets (both rocky ones, and gas giants), moons, asteroids, comets, and tiny fragments.

Some of those tiny fragments are swept up by the Earth in its orbit around the Sun. If they just flash briefly across the night sky, they are called "meteors." If they are large enough to not be completely burned by entering the Earth's atmosphere and so land on the surface, they are called "meteorites." These meteorites potentially contain a great deal of information about our primitive solar nebula, its composition, and formation. A few have even been blasted from the surface of Mars by huge impacts there, bringing us also knowledge of our neighboring planet in the form of meteorites.

Some scientists and others favor a theory called "panspermia," in which comets and meteorites have brought the first life forms to Earth. The actual origin of life, whether on Earth or beyond, is a question for the relatively new science of astrobiology. It is becoming clear from studies on Earth, that life can form and thrive in environments that we would find impossible. These harbor the "extremophiles," like the sea creatures we find next to deep-water ocean vents.

Part of astrobiology is the quest for planets that are sufficiently like our solid Earth and correctly positioned in relation to a suitable star, to favor the support of life (see the introduction to part 3, section C, and Script #9). The first planets to be found around stars beyond our solar system were the "hot Jupiters," that is, gas giant planets close to their host stars. This was due to an observation and selection effect, since they were the most easily detected by astronomers from the wobble of their host stars. However, the Kepler space telescope had sufficient sensitivity to track the tiny variation in a star's light while a planet transited across its face. It was therefore possible to discover stars with planets more like our Earth, and at distances from their host stars where water could exist in liquid form. After the data from Kepler, astronomers have learned that the pattern of planets around other suns is more like the pattern in our own solar system. We are beginning to characterize these Earth-like planets (see Script #9), but identifying which ones may harbor intelligent life is still a guessing game.

Space Science and Astronomy Script Package #10
"Early Humans at Herto and the Discovery of
Wandering Stars, 160,000 Years Ago"

Lessons in History and Culture

Firm evidence now exists in the archaeological record that members of the genus *Homo* and species *sapiens* were living in East Africa, in present-day Ethiopia, around 160,000 years ago. Artist reconstructions of the skulls and post-cranial remains reveal an individual remarkably like modern man. However, it was not modern man, but a near relative named *Homo sapiens idaltu* (White 2003). According to statistical analysis of this hominin's bones, he did not fall anatomically into the skeletal ranges of modern man. Still, he was an active hunter of hippopotamuses, and he engaged in a kind of post-mortem mortuary practice that suggests some kind of ritual. Was this what we would call "religion"? No one is yet willing to hazard a guess, but there is a strong likelihood that this sub-species of *Homo sapiens* did engage in very basic "religious thought."

The script at Herto shows two individuals, a male hunter and a female approaching marriageable age. The man, "Abo"—the word for "grandfather" in the local Ethiopian dialect—is wooing the woman with food. He offers her snails and mussels, and then, he promises roasted hippo meat. It is a very gentle, tender courtship, and he ends up attracting her not only with food, but with ideas, especially ideas about the stars and how he has noticed that some of the stars "wander." She is intrigued.

"El" is named after "L," which is the first recorded haplotype according to research on mitochondria in the cells of living humans. Fancifully, the character of "El" in this script could be thought of as "Mitochondrial Eve," the woman whose mitochondrial DNA, or later versions of it, shows up in all living humans today. All types of such DNA coalesce at about 200,000 years ago, suggesting that the woman who gave rise to us all lived in East Africa around that time.

Lessons in Space Science and Astronomy

The difference between the relatively fixed stars that appear in star charts and "wandering stars," or planets, has been described in the introduction just above to this part 3(D). The paths of the planets, as they would be seen from a vantage point above the solar system, are ellipses with the Sun at one of their foci. When planets are seen from Earth, "wandering" among the fixed stars, they can be puzzling until the geometry is understood. Then, the reasons become clear why only Mercury and Venus show phases, like the Moon, and stay close to the horizon, while other planets go backwards for a time in their normal west-to-east progress among the stars (retrograde motion).

The male character in this script mentions a "bluish star" that is clearly a planet rather than a star, and most likely, Venus. Its opaque clouds of sulphuric acid are highly reflective and completely cover the planet's surface, making Venus appear quite blue, as well as the brightest object in the night sky other than the Moon. The "red star" going one way and then the other against the background of fixed stars, is Mars.

Worksheets for Use before a Script

I. Keywords

Colors of stars	"Mitochondrial Eve"
Haplotype	"The Five Visible Planets"
Homo sapiens idaltu	Star chart
Hominin	Yardi Lake
Mitochondrial DNA, or mtDNA	

II. Questions for Investigation Before the Script

1. Who was the researcher who named *Homo sapiens idaltu*, and what is the approximate time period in which this hominin roamed East Africa?

2. Is it likely that *Homo sapiens idaltu* had language? Why or why not? Does language ability have any bearing on this hominin's likelihood of noticing star positions and movements? Explain.

Discussion Questions for Use After the Script

1. The male character in this script mentions a "bluish star" whose movements are different from the others. What is this star and how is it different?
2. What is the "red star" that goes one way and then the other? Why does its "wandering" differ from the bluish star.

Costumes and Props for Inexpensive Productions

For this script, just as for Script #7 on Early Humans at Kenya Cave, animal skins are appropriate since nighttime temperatures can dip uncomfortably in Ethiopia. Fake fur can be found that is relatively inexpensive, and pieces used to suggest animal skins. They can be draped over skin-colored pants and shirts, tights, leotard, or a catsuit.

Introduction to a Script on Early Man's Probable Discovery of 'Wandering Stars' (Planets)

The tension in this script derives from the subtle maneuvering of Abo, who is essentially making a marriage proposal to a coy young woman, El. The dialogue shows that each is sensitive to the persuasion of the other. It is the pleasant tension of romance and courtship, which is a very necessary sort of tension that sets these two characters up to marry and have children.

Script
"Early Humans at Herto and the Discovery of Wandering Stars, 160,000 Years Ago"

Characters:

EL, a young woman nearing marriageable age. She is named after "Mitochondrial Eve," who founded haplotype "L" (so, "El"), according to genetics research on mitochondrial DNA. The mitochondrial DNA of every living human is descended from hers.

ABO, an adult male, a hunter. His name means "grandfather" in the local Saho-Afar language of Ethiopia.

Place: Exterior.

Location: A man and a woman sit by an East African lake in the evening, about 160,000 years ago.

Time: Early evening, before the sun has set.

A [SLIDE OR SIGN] gives the year, place, and time:

"160,000 Years ago, alongside ancient Yardi Lake, in present-day Ethiopia, East Africa."

ABO

Come, let's sit here beside the lake, and roast these mussels and snails. They're good eating!

EL

She smiles.

You are clever with your gifts, Abo.

ABO

I'm not so clever, just determined.

EL

Are you as determined as that mother hippo who was trying to protect her calf? That was quite a battle today! I heard the noise from as far as our camp site!

ABO

Ah, well... She gave us a good fight—as well she had a right. It took the strength of four men to bring her down, and the calf was a fine bonus. It has tender meat. I will bring you some.

EL

You see—I said you were clever with your gifts.

ABO

And I said I was determined. There's a difference. It took cleverness to bring down the hippo, and strength! But it takes determination for what I want.

He laughs aloud.

EL

She smiles shyly.

You have to hold back, you mean, don't you? And you don't have to hold back with the hippo! Right?

ABO

Joking, with mock gravity.

I tell you what. It takes determination to skewer these little shells—the mussels from the lake and the snails from the reeds, and to wait until they are roasted just so—until they are tender and fragrant, and then wait... until they are... ready... to... plop into your mouth!

Laughing, he places a morsel in her mouth.

EL

She laughs heartily.

Abo... You are a jokester, Abo. You make me laugh!

ABO

Look up, El. See that star? The very bright one.

EL

She looks up, mouth agape.

The one that's bluish, is that the one?

ABO

He pops another morsel in her mouth.

Ah-ha! ...Yes, the bluish one. It's different from the rest—not all the rest, but most of the rest.

EL

And how do you know this, Abo?

ABO

I've been looking carefully at them for a long time. I come here by the lake, and I look up every night. I see their positions, and compare them the next night.

He points.

The red one—right there—is different from the rest. It goes one way, then it goes the other way. It can't make up its mind! It's not like the other stars, night after night. You understand?

EL

Yes, I think. Are there other stars like that—who wander around like a wildebeest or an antelope? They wander all around looking for good grassland. Maybe that's what the red and the blue wanderer are doing—looking for something good to eat!

ABO

That's an idea. I hadn't thought of it. Maybe. There are some others that do the same.

EL

Why are you so interested in the stars, Abo?

ABO

I'm not sure... But, it seems to me that some things here on the land happen when the stars are just so. It seems to me there is a connection.

EL

Wide-eyed.

I'm beginning to see! It's a way to tell what will happen...

ABO

Yes, the stars can foretell much, but not everything. Some things we have to work out, ourselves, down here on the land. ...And so, I will meet you here again, tomorrow evening, El, and I'll bring some roasted baby hippo meat for you.

EL

That's very kind of you, Abo.

ABO

And then, one moon from now, I will go and talk with your father. Would that be okay with you?

EL

She nods and smiles, looking down.

Yes, that would be fine with me. And then, you can teach me even more about the stars, Abo.

Space Science and Astronomy Script Package #11
"Science and Discovery of Biosignature Data from Exoplanets Nearest to Earth, A.D. 2075"

Lessons in History and Culture

This script portrays two professional individuals—an astrobiologist and an engineer—working together to receive and analyze data on biosignatures from one of the Kepler exoplanets. Unlike the Script #9 on the starship to an exoplanet, which takes place four centuries after this near-future date (A.D. 2075), the two individuals are not culturally or ethnically distinct according to their names, clothing, or language. They act out their roles in a locale for serious scientific research: the earthside control room for an orbital telescope, which has been collecting data on one of the Kepler exoplanets. They are physically located at the Institute of Astronomy at the University of Hawai'i, in Hilo, which is a real place today. They are not at a high-elevation facility such as the ESO Very Large Telescope, which lies high in the Andes Mountains, in Script #2. There is no need to be.

The only outstanding feature of the script from the viewpoint of history and culture is the "scientific culture," in which both observers are operating. A good project for a student might be to characterize this "occupational culture." Features of the culture are rapidly spreading as increasing numbers of trained professionals travel more and more to meet in increasingly far-flung locations, in order to learn about new science and to participate in their professional associations.

Lessons in Space Science and Astronomy

The focus of Script #11 is the detection of chlorophyll and other biosignatures on exoplanets. The science in the script is based partly on xray crystallography (knowledge of diffraction effects) and partly on spectroscopy. We speculate that the characters in the script are using a new technology, i.e., the extension of resonant inelastic xray scattering (RIXS), which gives xray spectroscopic information on the geometric structure of organic molecules. Our speculation hinges upon being able to detect the xrays that are reflected into space when a star irradiates an exoplanet's surface (as solar flares do onto Earth). Our speculation then relies on the RIXS produced in this way making it possible to detect very distant bio-samples on an exoplanet. Regardless of our inability to detect such differences from an exoplanet now, in the centers of the rings of the hemoglobin complex and the chlorophyll molecule the interchange of the Fe^{II} and Mg^{II} atoms is very real, as shown in Figure 7.

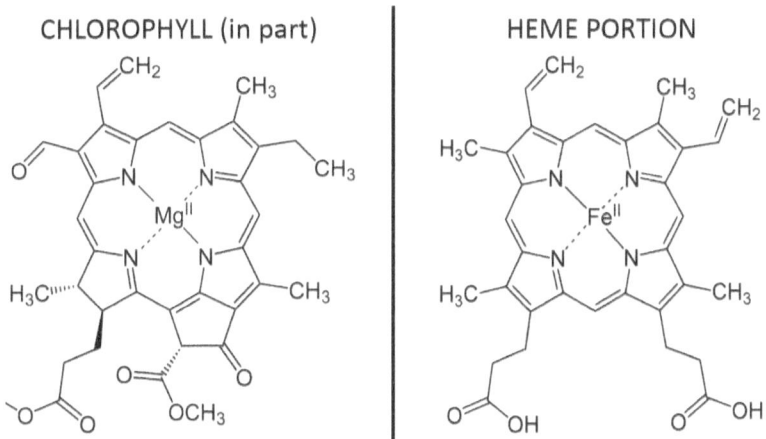

Figure 7. Images of a Chlorophyll Molecule and the "Heme" Portion of the Hemoglobin Complex. (Credit for each component: Yikrazuul)

The Keywords, whose number could be best trimmed or divided among a class, make it clear that there are many science lessons underlying this script. One important lesson is the power of spectroscopy, which may enable astronomers to learn much about life beyond the Earth.

Worksheets for Use before a Script

I. Keywords

Astrobiologist

Biosignature

Chlorophyll

Diffraction pattern (for organic molecules)

Electron density mesh map

Exoplanet; Kepler sample of exoplanets

Expert program; expert system

Futurist

Hemoglobin

Jules Verne

Orbital station

Packet of data, or data packet

Pharmaceutical

Protein crystal structure

Solar flare

Spectroscopy; spectroscopist

Spectrum (light)

The Red Edge

Xray crystallography

II. Questions for Investigation Before the Script

1. What is the Kepler sample of exoplanets?
2. What are some of the earliest biosignatures used by astronomers?
3. What "ring" do the chlorophyll molecule and hemoglobin complex have in common?
4. Does the field of "xray spectroscopy" exist today? What might it accomplish in distinguishing organic molecule structure?

Discussion Questions for Use After the Script

1. Solar flares might interfere with data transmission from a satellite or orbital station. Why?
2. Why did the two scientists in the script think that BITSSI was sending the same thing twice?

3. Why might a pharmaceutical researcher be suited for research on biosignatures?
4. These two researchers were working in Hawaii, but not at the location of a telescope. Why would some locations in Hawaii be a good location for a telescope?
5. These two researchers had something similar in their occupational histories. What was it, and how did it affect their work?

Costumes and Props for Inexpensive Productions

The authors took a very simple approach to costuming for Script #11. Each one wore a Hawaiian shirt with parrots, and informal pants. The garb for most astronomers and engineers working in control rooms is jeans and a shirt, perhaps a sweater if at high altitude.

The script would play very well against a backdrop slide of an orbital telescope control room. However, most essential is a series of visuals (or slides of some type) showing the data being downloaded by BITSSI. Gradually, from top to bottom, "she" (the telescope) reveals the two similar chemical species being identified by the characters, Jean and Jeff, and more important, how the molecules differ (Figure 7).

Introduction to a Script on the Science of Investigating Biosignature Data on Exoplanets

There are two sources of tension in the script. The first concerns the unpredictable and problematic nature of satellite-to-Earth communications, and the scientists' instrumentation—especially in an environment of sunspots. The characters lose contact with their orbital telescope for a time, but eventually, they are able to receive the large amount of data that BITSSI is trying to send them.

The second source of tension is the lead-up to the final reveal, in which two images are sent by the telescope (Figure 7). At first, the scientists think they are identical images, but they eventually determine that they are not. Students should note when the astrobiologist becomes convinced they are actually two different images, and why she concludes this.

Both students and teachers may wonder about the existing protocol that could follow the identification of an "intelligent signal," as by SETI (the organization that searches for extra-terrestrial intelligence). If a researcher finds an intelligent signal, do they have to notify anyone? The answer according to research by these two authors is, yes, they do. They have to notify the President. There are also United Nations protocols for notifications in case intelligent life is located. However, if the signal only demonstrates chlorophyll, the rules are unclear. If the data show something pointing to complex life, such as hemoglobin, there is no clear guidance either.

Script
"Science and Discovery of Biosignature Data from
Exoplanets Nearest to Earth, A.D. 2075"

Characters:
JEAN, Astrobiologist, Principal Investigator, ATFIRST Biosignatures Project
JEFF, Chief Engineer, ATFIRST Biosignatures Project

Place: Interior.
Location: Earthside Control Room for BITSSI, the orbital Binational Interstellar Telescope for Space Science Investigations, Institute for Astronomy, University of Hawai'i, Hilo.
Time: 2 a.m.

A [SLIDE OR SIGN] gives the year, place, and time:

"A.D. 2075, November 2. Earthside Control Room for BITSSI, the orbital

Binational Interstellar Telescope for Space Science Investigations, at the Institute for Astronomy, University of Hawai'i, Hilo. 2 a.m."

JEFF

At a console, dressed in a Hawaiian shirt.

BITSSI's cranky tonight. Her signal from the orbital station is fluctuating.

JEAN

At another console, in another Hawaiian shirt.

Don't tell me that, Jeff! Not tonight! Tonight's the big reveal! ...at least for Kepler 517 b...

JEFF

I think we've still got interference from last week's solar flares.

JEAN

We need a clean line. That packet of data is enormous! We need as few errors as possible to reconstruct the diffraction pattern.

JEFF

Transmission's down!

JEAN

Completely?

JEFF

Down.

JEAN

Sitting back.

We'll wait a couple hours, then try tomorrow.

JEFF

Huffing, he also sits back.

I can't believe we're still working with that Kepler sample of planets!

JEAN

They're like old friends, Jeff, and they're the best we've got. They're relatively close to Earth and they still have the best chance of showing biosignatures. And remember, two of them had *The Red Edge!* ...Drumbeat! I remember how excited all the astronomers were—and I wasn't even in the field yet. It was big news ten years ago!

JEFF

You know, I never asked you how you got into this project. You're a biologist, aren't you? You were in drug research, right?

JEAN

I was. I was in pharmaceutical research, trying to find a big, bad molecule that would stop the cancer that killed my Dad. I was on a mission!

JEFF

Why the change? I don't understand how you got into the ATFIRST project...

JEAN

Well, *at first*, so to speak, they had to convince me they were for real. I thought they were crazy. NASA comes to me looking for a researcher in xray crystallography. I said, "I'm used to looking down. You want me to look up?"

JEFF

He laughs.

I had the same experience, in a way. I cut my teeth on high-speed, dedicated computer networks—here on Earth! It took a while for me to re-orient *upward!*

JEAN

...But they were right, and they teamed me up with the perfect person, the very best spectroscopist... Anyway, together, we worked out this new understanding about the relationships between the output from xray crystallography—what I was doing—and xray spectroscopy—which was what he was doing.

JEFF

So this wasn't your idea?

JEAN

No indeed. Some *wild futurist* at NASA dreamed this up! The crazy thing is—it just might work! Jules Verne all over again!

JEFF

Yeah, it just might work... That's the weird thing.

JEAN

It *might.* We were going to see tonight in this run from Kepler 517 b. BITSSI's been collecting data for us, for the past four months. She's been busy! But, we may have to wait.

JEFF

Back to your story. So, you figured out how to get a diffraction pattern *from a spectrum?*

JEAN

Think about it. A diffraction pattern and a spectrum are both two-dimensional representations of three dimensional molecules. There are correspondences between signals in a spectrum and the structure of my big molecules. We just had to investigate what those connections were, and see if you could predict one from the other.

JEFF

You mean, if you could you get the shape of a molecule from a spectrum...

JEAN

Or, a lot of spectra, as it turned out! Anyway...it was a whole new field ten years ago—

JEFF

—BITSSI's back up!

JEAN

Do say?

JEFF

Strong signal. Good. Let's look at the data coming in...

JEAN

She comes over to his console and looks on.

JEFF

Two data streams. Two images. That's odd.

JEAN

Hmm. I forgot to tell you that we taught
her how to learn.

JEFF

BITSSI's an AI, too?

JEAN

Well, she has an expert program as part
of her decision-making—

JEFF

—She's an AI!

JEAN

Forgot to tell you that—Sorry! ...I mean,
we didn't necessarily expect just one
molecule from an entire exoplanet! ...So,
if BITSSI's collecting multiple spectra and
assembling an electron-density mesh map
for a 3-D molecule, and, she sees that there
can only reasonably be two structures,

she splits the output... I wasn't expecting two, but I'll take two. It's reasonable. 'Course, she may be wrong and they're really the same big molecule! ...Maybe, we'll see. ...And hopefully, it's chlorophyll.

JEFF

Jean, if we've got chlorophyll, we need to call the press.

JEAN

Indeed, but let's make sure. ...Okay, you can see the protein crystal structures she's beginning to assemble... Interesting...

JEFF

Looks like chlorophyll to me...

JEAN

Indeed, it does, but so does the other one! Okay, so BITSSI's fallible. Good to know when your machines can make mistakes...

JEFF

The images are diverging, Jean.

JEAN

Hmm.

JEFF

What's the other one? It looks like chlorophyll at first, but now it's not the same. Maybe they've got some weird chlorophyll on Kepler 517 b! ...Or *worms with green blood!* I remember that from my high school biology...

JEAN

She straightens up and looks away from Jeff's screen. She places a hand to her forehead.

The other one's not chlorophyll, Jeff. It's different in some very specific ways, although the outline is similar.

JEFF

What is it?

JEAN

It's the "heme" portion of hemoglobin, Jeff. See the dense region in the center that's so different. That's iron. We've probably got evidence of a hemoglobin molecule on Kepler 517 b.

She sits down.

JEFF

Hemoglobin?

JEAN

Hemoglobin.

Space Science and Astronomy Script Package #12
"Miners in the Asteroid Belt, A.D. 3021: New Occupations and an Old Culture"

<u>Lessons in History and Culture</u>

This script takes place in A.D. 3021, further into the future than any other script, except Script #3. However, it envisions a very old industry—mining—and a conversation between a mining administrator and a miner who is back from spending about a year away. As the story of the miner unfolds in the script, it becomes obvious that he escaped a close call with death, and that someone decided *not* to try an attempt to rescue him. Therefore, the exchange between him and his boss is full of more anger than any script in this volume. The authors found that the audience is quite sympathetic to his plight. He's glad to be back, but he wants to know who "betrayed" him and left him to die on the asteroid he was mining.

As long as there is mining, there will be the occupational types portrayed in this script. The miner is a risk-taker. He knows it and he likes "the life." The administrator has the marks of greed, and keeps her eye clearly on the "bottom line," even when conversing with this miner who has almost literally "returned from the dead." Yet, the characters are more complex than that. The administrator has her soft side, too, and Jacko, the miner, has his inexplicable fondness for an occupation "as old as the hills"—only, in this case, he mines asteroids.

The mining of asteroids has been the lore of science fiction, video games, and futures research until very recently. Now, it is a respectable business venture, and may be a very profitable one that avoids transporting precious materials, including water, off the Earth, for construction projects in space and on the Moon. It is far more economical to have iron, trace metals, water, and even precious metals for delicate electronics already in Earth orbit (or otherwise accessible), rather than spending money and energy to lift them off Earth.

Lessons in Space Science and Astronomy

It will not be a surprise that asteroids are not all the same. Different types reflect the processes of solar system formation, so their classification has been an important tool in understanding their origins. Classification of asteroids will have a new importance when mining them becomes a commercial reality.

This script is located on Iphigenia, an asteroid in the main asteroid belt between Mars and Jupiter. Each main belt asteroid has an independent orbit around the Sun. The time taken for one orbit—the asteroid's "year"—depends on its distance from the Sun, and its path will depend on the eccentricity and inclination of its elliptical orbit. Asteroids with different distances, eccentricities, and inclinations, that are starting their individual trips around the Sun will get close again—will be "in conjunction"—in different fractions of their years, depending on their respective orbital parameters. In this script, we imagine that Iphigenia and the mine jockey's asteroid come back into conjunction in a reasonable time—an Earth year. The geometry of orbits and conjunctions can be more easily understood by drawing them.

Worksheets for Use before a Script

I. Keywords

Artificial day	Iphigenia (the character, and the planetary body)
Asteroid	Synodic period
Conjunction	S-type asteroid

II. Questions for Investigation Before the Script

1. When two planetary bodies (including asteroids) are "at conjunction," what does that mean?

2. What are the basic types of asteroids, as they are presently classified? What is the main factor that distinguishes different types of asteroids?

Discussion Questions for Use After the Script
1. "Positional astronomy" is important for understanding the tension in this script. Describe the relative positions of the Iphigenia asteroid and the asteroid on which Jacko was mining when he lost communications with his base. What can you determine from the dialogue?
2. Jacko used the term, "mine jockey," to describe himself. Do you think this is a bona fide new occupation, or not? Have there been similar occupations on Earth in the past?
3. Jacko's emotions range widely in this script. How did he end up seeing his long journey around the Sun, and how did he feel about it?
4. How do the shapes of asteroid orbits vary?

Costumes and Props for Inexpensive Productions
The authors obtained a child's toy "mining hat" with a light on the front, from an online resource, at a modest price. The hat for the administrator in the script was an anonymous pilot's cap, also from an online source at low cost. These were the only clothes or props used. The trickiest part of this script is the re-creation of low gravity in an office setting, because the asteroid 112 Iphigenia (nicknamed "Iffy" in the script) is a relatively small planetary body that does not have much gravity. Student productions could experiment with this difference, imagining a particular fluid gait, or, simply proposing that the characters wore weighted shoes.

The script will play well to a backdrop of a map of Iphigenia's position, and then, the position of the asteroid on which Jacko had to spend a long time in hibernation until it came back into conjunction with Iphigenia. The astronomy lesson can be illustrated on slides, while the two characters speak. A slide with futuristic images of the two characters has also served the authors well.

Introduction to a Script on Asteroid Mining

Jacko's anger is more difficult for inexperienced actors to convey than often supposed, because it is forceful, sometimes loud, and it needs to "sound true" to the audience. Jacko's sarcasm is easier to portray, as he mocks and chides his boss about her failure to rescue him. In the end, there is a reconciliation, but with a surprise twist at the end.

There is tension throughout the script, and it is important for the action to have its quieter moments for contrast, as when the administrator shows real regret and concern. The forcefulness of the language cannot be shrill throughout the script, only at its most intense. The actors must modulate their voices for maximum effect.

Script
"Miners in the Asteroid Belt, A.D. 3021: New Occupations and an Old Culture"

Characters:
BERNADETTE Desjardins, Administrator, Main Belt Mining (MBM), on Asteroid 112 Iphigenia
JACKO Slack, Mine Jockey, recently rescued from a mining operation on an S-Type asteroid (with platinum and gold), after a one-Earth-year absence.

Place: Interior, Admin Dome, on the asteroid 112 Iphigenia ("Iffy").
Location: A man enters a woman's office, and makes himself comfortable.
Time: 10 o'clock, Day Phase.

A [SLIDE OR SIGN] gives the year, place, and time:

"A.D. 3021, April 18. Admin Dome on 112 Iphigenia, Main Asteroid Belt. 10 o'clock, in artificial day."

JACKO

He crabwalks into her office, "sits" in low gravity, and comfortably slouches.

Bern-a-*dette...* Bernadette, my luv...

BERNADETTE

Angry.

How dare you! How dare you address me that way! I am your administrator and I am 'Administrator Desjardins' to you!

JACKO

Aaoow.... I've made the lady *angry.* So sorry, luv, but you didn't seem happy to see me...

BERNADETTE

Contrite.

Of course I'm happy to see you, Jacko. We all are. We just don't know how you managed to live on that S-type hunk of rock for *a whole Earth year!* How did you survive that trip around the sun? You couldn't have had enough oxygen...or nourishment...

JACKO

Impressive, wasn't it? I'm *impressed* by myself, I am, I am!

BERNADETTE

How did you *do* it, Jacko? By the time we figured you were in trouble...you were too far away from Iffy's orbit, for us to help!

...Did you get any platinum, by the way?

JACKO

Always with her eye on the bottom line... Well, I did manage to cop a hundred kilo or so...

BERNADETTE

A hundred kilo of platinum? That's impossible!

JACKO

Aaoow... I'm exaggerating just a bit, luv, but almost. I was there a *long time*, you recall!

BERNADETTE

She looks down, then up at him, puzzled.

Seriously, Jacko, how did you stay alive for the entire Earth year it took that asteroid to come back into conjunction with us? You had no communications that we could detect...

JACKO

Skeptical.

You looked for me, did you, luv? That's sweet... because *my comm equipment was receiving long after I couldn't send any more!* I got Earth traffic—*Earth traffic!*—long after I heard from Admin here on 112 Iffy... Anyway, I made it with the gas and equipment I *stole from Supply!* You think I'm nuts? I wouldn't get caught out there without my *equipment!*

And then, like a good lad, I put myself to sleep, hibernating like, hoping the gas I had would take me around old Sol!

BERNADETTE

Amazing... But Jacko, we looked for your signal. I pleaded with Earth to let me send a rescue ship for you, but they wouldn't agree to it...

JACKO

Still skeptical.

You decided not to come for me, didn't you? Why? That's what I got to know. *Why?*

BERNADETTE

They wouldn't spring for the fuel, Jacko. They said it would cost too much to get you. They said you knew the risks... I'm so sorry... We never thought you were still alive—

JACKO

Angry.

—made it, *at conjunction,* and show up here alive and all, making the likes of you *feel guilty!* Who made the decision, Bernadette, that's what I gotta know. *Who made the decision?*

BERNADETTE

I'm sorry, Jacko. I tried, but we were short on people and we hadn't made our quota in metals, and it was going to be a trip for someone who was probably already dead, so I—

JACKO

—*You!* You betrayed me, Bernadette! *You left me there to die!*

BERNADETTE

She wipes away a tear.

I'm sorry, Jacko. I made the best decision I could at the time. ...I'd make it again.

JACKO

Calmer, he sits back.

That's me girl. That's alright. I just need to know when my next assignment is, Bernadette.

BERNADETTE

She looks up quickly, with an unbelieving expression.

You're going out *again?* After all you've been through?

JACKO

I'm a *Mine Jockey*, Bernadette, I ride them asteroids like a bucking bronco, just out of the Old West! Right? It's what I was meant to do! I'm a *Mine Jockey!*

REFERENCES AND RECOMMENDATIONS

Alcubierre, Miguel. 1994. The Warp Drive: Hyper-Fast Travel within General Relativity. *Classical and Quantum Gravity* 11: L73-L77.

American Astronomical Society. 2005 *A New Universe to Discover: A Guide to Careers in Astronomy.* Washington DC: American Astronomical Society. Accessed February 9, 2017, https://aas.org/files/resources/Careers-in-Astronomy.pdf

Armstrong, Mabel. 2008. *Women Astronomers: Reaching for the Stars.* Marcola: Stone Pine Press.

Corbally, C. J. and Rappaport M. B. 2013. Crossing the Latest Line: The Evolution of Religious Thought as a Component of Human Sentience. *Evolution: Development within Big History, Evolutionary and World-System Paradigms.* Yearbook. Eds. L. E. Grinin and A. V. Korotayev. Volgograd: Uchitel. Pp. 197-218.

Davies, Paul. 1995. *Are We Alone?: Philosophical Implications of the Discovery of Extraterrestrial Life.* New York: Basic Books.

Dean, L. G., R. L. Kendal, *et al.* 2012. Identification of the Social and Cognitive Processes Underlying Human Cumulative Culture. *Science 335:* 1114-1118.

Dorsey, Gary. 2000. *Silicon Sky: How One Small Start-up Went Over the Top to Beat the Big Boys Into Satellite Heaven.* New York: Basic Books.

Easton, Thomas. 2004. *Careers in Science.* New York: McGraw-Hill.

Elvis, Martin. 2014. The case for applied astronomy. *News and Reviews in Astronomy & Geophysics* 55(1): 11-12.

Fohlmeister, Janine, and Christiane Helling. 2014. Careers in Astronomy in Germany and the UK. *Astronomy & Geophysics* 55: 2.31-2.37.

Freedman, Jeri. 2012. *Your Career in the Air Force.* New York: Rosen Publishing Group.

Hutson, Matt. 2007. *Totally Amazing Careers in Aerospace.* San Diego: Sally Ride Science.

Hutson, Matt. 2007. *Totally Amazing Careers in Engineering.* San Diego: Sally Ride Science.

Infobase Publishing. 2008. *Discovering Careers for Your Future: Space Exploration.* New York.

Institute for Career Research. 2007. *Careers in Aerospace Engineering: Aeronautics—Astronautics.* Chicago: www.careers-internet.org.

Institute for Career Research. 2006. *Careers in US Air Force.* Chicago: www.careers-internet.org.

Ivey, Catherine. 2007. *Totally Amazing Careers in Space Sciences.* San Diego: Sally Ride Science.

Ivey, Catherine Lee. 2010. *Cool Careers in Space Sciences.* San Diego: Sally Ride Science.

Koenig, Robert. 2009. Minority retention rates in science are sore spot for most universities. *Science* 324: 1386-87.

Kranz, Gene. 2001. *Failure Is Not an Option: Mission Control from Mercury to Apollo 13 and Beyond.* New York: Simon & Schuster.

McDougall, Ian, Francis H. Brown, and John G. Fleagle. 2005. Stratigraphic Placement and Age of Modern Humans from Kibish, Ethiopia. *Nature* 433, 733-736.

McGonigal, Jane. 2011. *Reality Is Broken: Why Games Make Us Better and How They Can Change the World.* New York: Penguin.

Mervis, Jeffrey. 2014. Studies suggest two-way street for science majors. *Science* 343: 125-26.

National Aeronautics and Space Administration. 2001. *Space-Based*

Astronomy: An Educator Guide with Activities for Science, Mathematics, and Technology Education. Houston.

National Research Council, Committee for a Decadal Survey of Astronomy and Astrophysics. 2010. *New Worlds, New Horizons in Astronomy and Astrophysics*. Washington, DC: National Academies Press (NRC 2010).

Rappaport, Margaret Boone, and Christopher Corbally. 2015. Matrix Thinking: An Adaptation at the Foundation of Human Science, Religion, and Art. *Zygon; Journal of Religion and Science* 50(1):84-112.

Stone, Tanya Lee. 2009. *Almost Astronauts: 13 Women Who Dared to Dream*. Somerville: Candlewick Press.

Teilhard de Chardin, Pierre. 2002. *The Heart of Matter*. Boston: Houghton Mifflin Harcourt.

White, Tim D., Berhane Asfaw, David DeGusta, Henry Gilbert, Gary D. Richards, Gen Suwa, and F. Clark Howell. 2003. Pleistocene *Homo sapiens* from Middle Awash, Ethiopia. *Nature* 423: 742-747.

ABOUT THE AUTHORS

Margaret Boone Rappaport, PhD, is a cultural anthropologist and biologist who works in the area of Science and Religion, and as a futurist, lecturer, and both fiction and non-fiction writer in Tucson, Arizona. As President, Policy Research Methods, Incorporated, Falls Church, Virginia, she was a contractor to federal and state agencies for over twenty years. She lectured in Sociology and Anthropology at Georgetown and George Washington Universities. She earned her doctorate at the Ohio State University in 1977. Her dissertation was on the adjustment of Cuban refugee women and families. Dr. Rappaport is a prize-winning short story and poetry writer.

Christopher J. Corbally, SJ, PhD, is a Jesuit priest and an astronomer with the Vatican Observatory Research Group, for which he has served as Vice Director, and liaison to its headquarters at Castel Gandolfo, Italy. He is an Adjunct Associate Astronomer at the Department of Astronomy, University of Arizona, and ministers to a wide variety of Catholics, including Native Americans, in Tucson, Arizona. He earned his doctorate in Astronomy at the University of Toronto in 1983. Dr. Corbally's interest in issues of science and faith is long-standing and well documented online. He is an active member and past-president of IRAS, the Institute on Religion in an Age of Science.

Rappaport and Corbally are Co-Founders of The Human Sentience Project LLC.

Web Site: http://TheHumanSentienceProject.org

www.ingramcontent.com/pod-product-compliance
Lightning Source LLC
Chambersburg PA
CBHW031944170526
45157CB00002B/386